Introduction to PSpice® Manual
Using Orcad® Release 9.2 for

Introductory Circuits
for Electrical and Computer Engineering

James W. Nilsson
Professor Emeritus, Iowa State University

Susan A. Riedel
Marquette University

PRENTICE HALL
Upper Saddle River, NJ 07458

Publisher: Tom Robbins
Supplement Editor: Jody McDonnell
Executive Managing Editor: Vince O'Brien
Managing Editor: David A. George
Production Editor: Barbara A. Till
Supplement Cover Manager: Paul Gourhan
Supplement Cover Designer: PM Workshop Inc.
Manufacturing Buyer: Ilene Kahn
Cover Art: Vanessa Piñeiro

 © 2002 by Prentice Hall
Prentice-Hall, Inc.
Upper Saddle River, NJ 07458

Printed in the United States of America

10 9 8 7 6 5 4 3 2 1

ISBN 0-13-067494-X

Pearson Education Ltd., *London*
Pearson Education Australia Pty. Ltd., *Sydney*
Pearson Education Singapore, Pte. Ltd.
Pearson Education North Asia Ltd., *Hong Kong*
Pearson Education Canada, Inc., *Toronto*
Pearson Educacíon de Mexico, S.A. de C.V.
Pearson Education—Japan, *Tokyo*
Pearson Education Malaysia, Pte. Ltd.

Contents

Preface

ABOUT THIS MANUAL

Introduction to PSpice expressly supports the use of OrCAD PSpice A/D, Release 9.2 (herein after referred to as PSpice) as part of an introductory course in electric circuit analysis based on the textbook *Introductory Circuits for Electrical and Computer Engineering.* This supplement focuses on three things: (1) learning to draw and simulate linear circuits using PSpice, (2) constructing circuit models of basic devices such as op amps and transformers, and (3) learning to challenge computer output data as a means of reinforcing confidence in simulation. Because PSpice is designed to simulate networks containing integrated circuit devices, its range of application goes well beyond the topics covered in the textbook. Even though we do not exploit the full power of PSpice, we begin the introduction to this widely used simulation program at a level that you can use to test the computer solutions.

The use of PSpice involves learning a new technical vocabulary and a number of specialized techniques. Hence, we designed this supplement to stand on its own as an instructional unit. Our decision to separate this material from the parent textbook also is a service to you: The portable format greatly facilitates your use of the supplement at a computer terminal. In this format, the supplement adds value to your study.

You may use PSpice to solve many of the textbook's Drill Exercises and Chapter Problems. Those Chapter Problems that we think are particularly suited to PSpice simulation are marked with a �P icon in the textbook. You are encourage to use PSpice to check your solutions to Chapter Problems, or to further explore the behavior of an interesting circuit.

Table 1: Relating Topics in this Supplement to Topics in the Text

MANUAL CHAPTER	TOPIC	TEXT CHAPTER
2–3	DC analysis	1–3
4	Thévenin Equivalents	3
4	Tolerance and Sensitivity	3
5	Op amps	4
6	Natural Response	5–6
7	Step response	5–6
7	Switches	5
8	Varying component values	6
9	AC Steady-state analysis	7
10	AC Power	10–11

INTEGRATING PSPICE INTO INTRODUCTORY CIRCUITS COURSES

Although some circuits courses cover PSpice as an independent topic, many instructors prefer to integrate computer solutions with the course. To support such integration topics appear in this supplement in the same order in which they are presented in the text. Table 1 summarizes the relationship between the supplement and the textbook.

ABOUT PSPICE

SPICE is a computer-aided simulation program that enables you to design a circuit and then simulate the design on a computer. SPICE is the acronym for a Simulation Program with Integrated Circuit Emphasis. The Electronics Research Laboratory of the University of California developed SPICE and made it available to the public in 1975.

Many different software packages are available that implement SPICE on personal computers or workstations. Among them, Orcad PSpice A/D, from Orcad, Inc., is the most popular. PSpice's popularity can be attributed to many factors, including its user-friendly interface, extensions to SPICE that support modeling of digital circuits and much more, and its no-cost basic version. This manual focuses on how to use Release 9.2, and all the examples were produced using this release. If you are using a different version of PSpice, or another package that implements SPICE, your interaction with the software may differ from what you see in this supplement.

We limited our applications of PSpice to the types of circuit problems discussed in the textbook. Although PSpice is a general-purpose program designed for a wide range of circuit simulation—including the simulation of nonlinear circuits, transmission lines, noise and distortion, digital circuits, and mixed digital and analog circuits—here we discuss the use of PSpice only for dc analysis, transient analysis, steady-state sinusoidal (ac) analysis, and Fourier series analysis. You should refer to the Orcad PSpice A/D User's Guide for information on all of the features of PSpice that are not discussed in this supplement.

We have included the Orcad Lite Version 9.2 in the back of this supplement. Insert the CD into your CD-ROM drive, and wait for the Orcad main menu to appear after a short animation. If the main menu is not displayed after one minute, choose the Start menu and enter `D:\setup.exe`, where "D" is the letter assigned to your CD-ROM drive.

Chapter 1

A FIRST LOOK AT PSPICE

The general procedure for using PSpice consists of three basic steps. In the first step, the user describes the circuit to be simulated or analyzed by drawing it in schematic form. In the second step, the user specifies the type of analysis desired, and directs PSpice to perform that analysis. In the third and final step, the user instructs the computer to print or plot the results of the analysis. In this first look, we will carefully develop a simple example to illustrate how to perform these three steps.

1.1   DRAWING THE CIRCUIT

To begin, run the OrCAD Capture CIS program, and select the option File/New/Project, as shown in Fig. 1. This option will invoke the New Project dialog box, as shown in Fig. 2. You should select a meaningful project name and the desired directory for your project files. You should also select the Analog Mixed-Mode Project Wizard, as shown in Fig. 2, which will guide you through the rest of the process.

You can create your PSpice project based on an existing project, or create a blank project. The dialog for making this choice is shown in Fig. 3. When you exit this dialog, you are back in Capture, but now have a circuit layout grid and the Capture toolbar, as shown in Fig. 4.

Now you are ready to draw your circuit. Begin by selecting the menu option Place/Part, or by clicking on the second vertical toolbar button. This will invoke the Place Part dialog. If you know the name of the library containing the part you want, highlighting the library name will display only the parts contained in that library. If the library containing the part

1

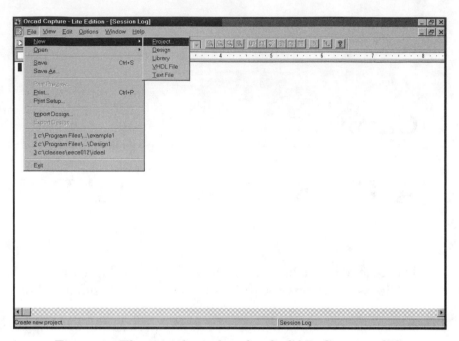

Figure 1: The initial window for OrCAD Capture CIS.

Figure 2: The New Project dialog box.

Figure 3: The dialog window for creating a PSpice project.

you need is not shown, you will need to add it by clicking on the Add Library button. The dialog for adding a library is shown in Fig. 5(a). Usually the libraries you need are in the PSpice subdirectory. As you can see in Fig. 5(b), we have selected the part named Vdc, a dc voltage source, from the library of sources named Source. If you are not sure of which library contains the part you need, you can highlight all of the libraries, and then all of the available parts will be displayed. When you click on a part name from the list its schematic is shown (see Fig. 5(b)), so you can verify that this is the part you want. Click on OK when you are ready to place the part in the schematic.

Figure 4: The circuit layout grid and Capture toolbar.

Figure 5: Selecting a part: (a) adding the library containing the part; (b) selecting the desired part.

Now you will be back in the Capture window and can place one or more copies of the part you have selected on the layout grid. You can change the orientation of the current part by typing Control-R. When you have finished placing as many copies of the current part as you need, right click on the mouse and select End Mode, as shown in Fig. 6.

Now you are ready to specify the attributes of the part (or parts) you have placed. Usually the only attribute you must specify is the value of the part, which in this case is the value of the voltage supplied by the dc voltage source. To change this value, double click on it, and you will see the

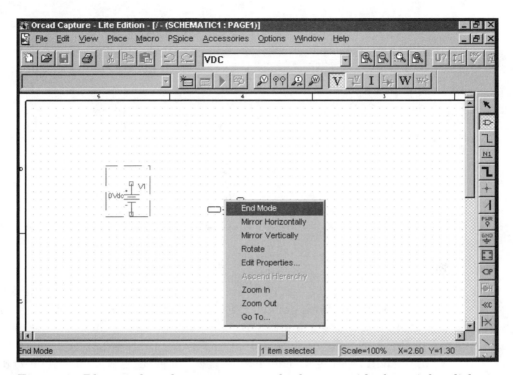

Figure 6: Place a dc voltage source on the layout grid, then right click on the mouse to finish with this part.

Display Properties dialog for this part, shown in Fig. 7. You can see from the figure that we have changed the value of the dc voltage source to 12V, or 12 volts. The number you specify may be an integer (4, 12, −8) or a real

Figure 7: Using the splay Properties dialog window to set the value of the dc voltage source.

Table 1.1: PSpice Scale Factors

SYMBOL	EXPONENTIAL FORM	VALUE
F (or f)	1E−15	10^{-15}
P (or p)	1E−12	10^{-12}
N (or n)	1E−9	10^{-9}
U (or u)	1E−6	10^{-6}
M (or m)	1E−3	10^{-3}
K (or k)	1E3	10^{3}
MEG (or meg)	1E6	10^{6}
G (or g)	1E9	10^{9}
T (or t)	1E12	10^{12}

number (2.5, 3.14159, −1.414). Integers and real numbers may be followed by either an integer exponent (7E−6, 2.136E3) or a symbolic scale factor (7U, 2.136k). Table 2 summarizes the symbolic scale factors used in PSpice and their corresponding exponential forms. Letters immediately following a number that are not scale factors are ignored, as are letters immediately following a scale factor. For example, 10, 10v, 10HZ, and 10A all represent the same number, as do 2.5m, 2.5MA, 2.5msec, and 2.5MOhms.

The process of locating a part, placing it on the layout grid, and setting the attribute values is repeated for each part in your circuit. Figure 8 shows the circuit we are working on, which has a 12 V dc source and two resistors with values of 3 kΩ and 6 kΩ.

One all of your parts have been selected, placed, and had their attributes specified, you are ready to wire the circuit. You can select the menu option Place/Wire, or click on the third vertical toolbar button. Wires are placed by clicking the mouse at the starting node for the wire, and clicking the mouse again at the ending node for the wire. Figure 9 shows the example circuit with all of the wires placed.

The last step in drawing the circuit is to add the circuit ground, which must be at the node numbered zero (0). The easiest way to do this is to click on the vertical toolbar button labeled GND, and select the part named 0, as shown in Fig. 10. (If you don't see this part, you will need to click the Add Library button and add the Source library from the PSpice subdirectory.) This is a very important step and easy to forget, but your circuit cannot be

Figure 8: All of the circuit components have been selected and placed on the layout grid.

simulated by PSpice unless it contains a ground at node 0! Place the ground component on the layout grid, and wire it to your circuit. The completed circuit is shown in Fig. 11.

Now we are ready to specify the type of analysis to perform on this circuit. Before we specify the type of analysis to perform, let's briefly examine what PSpice will do with the circuit schematic we just drew. In order for PSpice to understand the circuit we have described, the schematic must be translated into a collection of statements that identify the circuit components, their attributes, and their topological connections. This collection of statements is called a netlist, and it is written to the output file before analysis begins. Sometimes it is useful to look at the netlist, especially if there is an error in the schematic. Therefore, the Appendix includes a subset of the netlist syntax, for your reference.

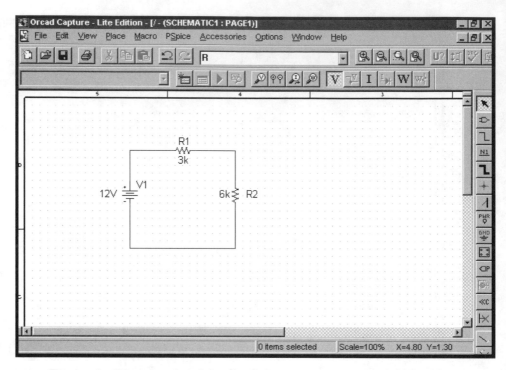

Figure 9: The circuit with all of the components wired together.

Figure 10: Selecting the ground component.

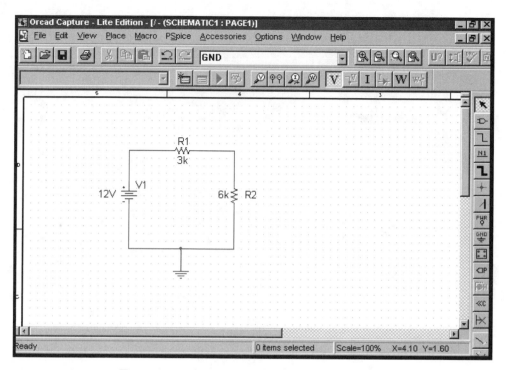

Figure 11: A completed circuit schematic.

1.2 SPECIFYING THE TYPE OF CIRCUIT ANALYSIS

You can ask PSpice to perform several different types of analysis on a circuit you have drawn. To begin, you need to select a name for the simulation. This is done by choosing the menu option PSpice/New Simulation Profile, which will generate the dialog shown in Fig. 12. We've chosen the name "bias" for this profile, as we will perform a simple dc analysis to determine

Figure 12: Creating a new simulation profile.

Figure 13: The Simulation Settings dialog window, used to specify the type of circuit analysis.

all of the node voltage values, a process known as "biasing". Click the Create button and the Simulation Settings dialog window will appear, as shown in Fig. 13. Use the Analysis Type dropdown list on the Analysis tab to select the desired type of analysis. We have selected Bias Point, as you can see from Fig. 13. You can also see in this figure that there are many other tabs in the dialog which can be used to further configure the analysis. Usually, the default values suffice, as they do in this example. Once you select the analysis type, click on OK to return to your schematic.

Circuit analysis begins when you select the menu option PSpice/Run. You will see the PSpice A/D window, as shown in Fig. 14. As the figure shows, this window is divided into three sections. In the lower left corner of the window is an output text section that displays the progress of the simulation. If errors were encountered in your circuit schematic or in the specification of the analysis, a brief error message will appear in this section, and simulation will not continue until the errors are corrected. In the lower right corner is a section that allows you to monitor the values of any parameters that may be changing as the simulation progresses. There are no such parameters in this simple bias example. Finally, the top section can be used for printing or plotting results of the simulation, the topic of the next section.

Figure 14: Circuit simulation using PSpice A/D.

1.3 SIMULATION RESULTS

Now we turn to the output from the simulation, which may be a text file, a plot, or both. In this simple example we cannot examine a plot, as no parameters have been varied. We examine the output from the bias analysis by clicking the third button on the left vertical toolbar. This opens the output file created during simulation in the top section of the PSpice A/D window, as shown in Fig. 15. Near the top of this output file is the netlist, the collections of statements that describe the circuit you represented as a schematic in a form that PSpice can understand. As you can see from the figure, the netlist consists of the three parts we drew: a 12 V dc voltage source between nodes 80 and 0 (the required ground node); a 3 kΩ resistor between nodes 80 and 110; and a 6 kΩ resistor between nodes 110 and 0.

If we scroll further down in the output file, we see the results of the bias analysis, including the value of the voltage at each of the non-ground nodes, as shown in Fig. 16(a). The voltage at node 80 is 12 V, which we expected since the 12 V dc voltage source is between node 80 and ground.

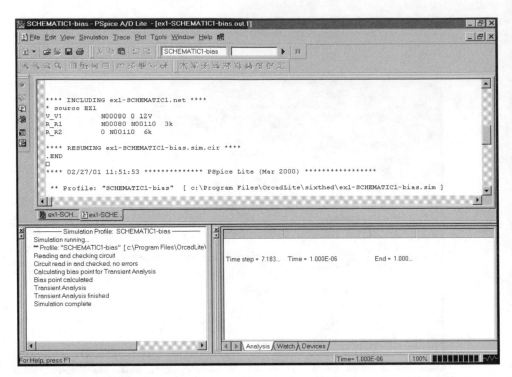

Figure 15: Displaying the output file produced during circuit simulation.

The voltage at node 110 is 8 V, which we also should have expected, since the circuit we created is a simple voltage divider and

$$v_{6k\Omega} = \frac{6}{3+6}(12) = 8 \text{ V}.$$

We can also see these node voltage values displayed directly on the schematic. To do this, exit PSpice A/D and return to the schematic. Click on the V symbol in the second-row toolbar, and the node voltages will be displayed, as seen in Fig. 16(b).

This completes the simple example. The remainder of this supplement describes the various types of components used in linear circuits, the other types of analysis available in PSpice, and the use of Probe to plot the results of circuit simulation.

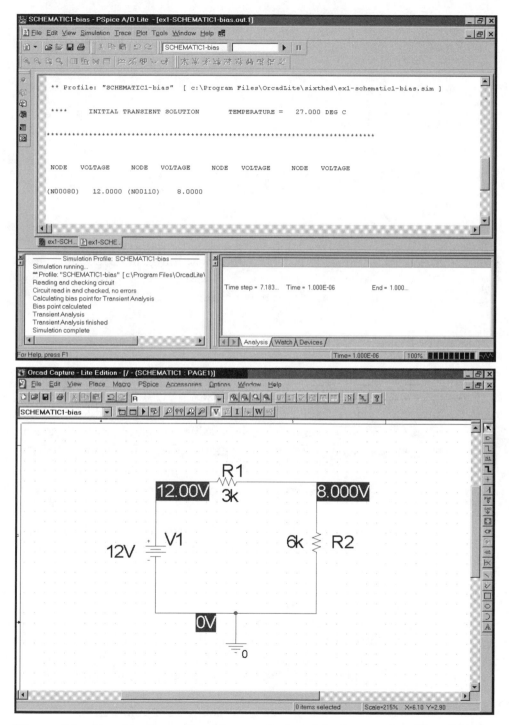

Figure 16: Results from the bias analysis: (a) the output file; (b) the schematic with node voltages displayed.

Chapter 2

SIMPLE DC CIRCUITS

In this chapter we look at circuits containing independent sources and/or dependent sources, together with resistors. We continue to analyze such circuits using a simple bias calculation.

2.1 INDEPENDENT DC SOURCES

Parts which model independent dc voltage and current sources are found in the Sources library, and are named Vdc and Idc, respectively. Once the independent sources are placed in the layout grid, double click on the value of the source to specify the desired value, which can be either a positive or a negative number. We saw an example of Vdc in Chapter 1.

2.2 DEPENDENT DC SOURCES

In discussing the dependent dc sources, we divide them into voltage-controlled sources and current-controlled sources. Placing and wiring dependent sources can be a little tricky, as they each require four wired connections — two that connect the source into the circuit, and two that connect the controlling voltage or current into the source.

We start with a voltage-controlled voltage source. A circuit containing this type of dependent source, together with an independent dc current source, is shown in Fig. 17. The part name for the voltage-controlled voltage source is E, and is found in the Analog library, as illustrated in the Place Part dialog for this part, shown in Fig. 18. The schematic representation of this part, as seen in the figure, has four connection points. The two points on the right, labeled + and − inside a circle, are used to wire this component into

Figure 17: Circuit used to illustrate a voltage-controlled voltage source.

Figure 18: The Place Part dialog for the voltage-controlled voltage source.

the circuit between the 6 A current source and the 4 Ω resistor (see Fig. 17). The other two nodes are attached to a circuit to measure the controlling voltage. These two nodes must be wired into the circuit in parallel with the component across which the controlling voltage is defined. This is the 3 Ω resistor in Fig. 17, where the controlling voltage v_Δ is defined.

The circuit schematic with all of the parts placed but none of the connecting wires, is shown in Fig. 19. Note that we have rotated the voltage-controlled voltage source in preparation to orient it for wiring. Before we wire the schematic, we need to specify the gain of the voltage-controlled voltage source. To do this, click on the schematic of the voltage-controlled voltage source (not its labels), which will highlight this part and place a

Figure 19: Parts placed for the schematic describing the circuit in Fig. 17.

dashed box around it. Then select the menu option Edit/Properties to invoke the Property Editor. Scroll horizontally in the property spreadsheet until you encounter the column labeled GAIN, double click on the value of the gain and change it to 3, as shown in Fig. 20.

Now the parts can be wired and the ground added to complete the schematic, which is shown in Fig. 21. Be careful to wire the circuit so as to achieve the polarities shown in the circuit diagram. You might need to zoom in on the schematic as you wire in order to see the polarity markings on the dependent source. Use the zoom toolbar button or the menu option View/Zoom.

The schematic representations for the remaining dependent dc sources are shown in Fig. 22. These dependent dc sources are also in the Analog library. The current-controlled current source has the part name F, the voltage-controlled current source has the part name G, and the current-controlled voltage source has the part name H. Remember to wire in the dependent current sources using the terminals with the arrow inside of a circle, shown on the right of the schematic for parts F and H. The dependent sources that are controlled by currents (parts F and G) have a large arrow

Figure 20: The Property Editor, used to change the gain of the voltage-controlled voltage source.

on the left side of the schematic (see Fig. 22). These terminals must be wired in series with the component through which the controlling current is defined. We will see an example of a circuit employing a current-controlled dependent source in the next section.

2.3 RESISTORS

Chapter 1 performed a dc bias analysis on a circuit containing resistors and a dc voltage source. There, we saw that the model of a resistor has the part name R, and this part is found in the Analog library. Once we placed the resistors in our schematic, double clicking on their resistance value allows us to modify the resistance. The following example includes the bias point analysis of a circuit containing an independent source, a dependent source, and some resistors.

Figure 21: The completed schematic for the circuit in Fig. 17.

EXAMPLE 1

a) Use PSpice to find the voltages v_a and v_b for the circuit shown in Fig. 23.

b) Use the PSpice solutions to calculate (1) the total power dissipated in the circuit, (2) the power supplied by the independent current source, and (3) the power supplied by the current-controlled voltage source.

SOLUTION

a) The circuit shown in Fig. 23 is represented as a schematic for PSpice as shown in Fig. 24. Note the wiring of the current-controlled voltage source. The pertinent PSpice output is shown in Fig. 25. From the PSpice output,

$$
\begin{aligned}
v_a &= \text{V(node 13)} = 104.00 \text{ V}; \\
v_b &= \text{V(node 20)} = 106.00 \text{ V}.
\end{aligned}
$$

Figure 22: The schematic for the current-controlled current source (part name F), the voltage-controlled current source (part name G), and the current-controlled voltage source (part name H).

Figure 23: Circuit for Example 1.

b) The total power dissipation value is automatically printed whenever you invoke simple dc analysis. In PSpice, this power represents the net power generated by the independent voltage sources in the circuit. If a dependent source or an independent current source is generating power, its value is not included. That explains why the PSpice output in Fig. 25 gives the total power dissipation as 0 W — there are no independent voltage sources in the circuit that was simulated. Therefore, to find the actual value of the power dissipated and supplied, we need to write the following equations:

Figure 24: The schematic representation of the circuit shown in Fig. 23.

1.

$$P_{5\Omega} = \frac{104^2}{5} = 2163.20 \text{ W};$$

$$P_{20\Omega} = \frac{104^2}{20} = 540.80 \text{ W};$$

$$P_{1\Omega} = \frac{(106 - 104)^2}{1} = 4.00 \text{ W};$$

$$\sum P_{\text{dis}} = 2163.2 + 540.8 + 4 = 2708 \text{ W}.$$

2. $P_{24\text{A}}(\text{supplied}) = 104(24) = 2496 \text{ W}.$

3. $P_{\text{H1}}(\text{supplied}) = 106(2) = 212 \text{ W}.$

Note that the sum of power dissipated equals the sum of power supplied.

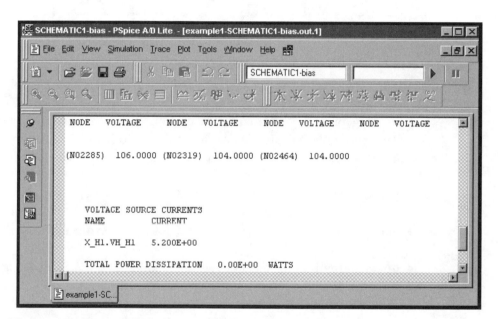

Figure 25: Part of the output file from the PSpice simulation of the circuit schematic shown in Fig. 24.

Chapter 3

DC SWEEP ANALYSIS

In the first two chapters, we instructed PSpice to compute the bias point (the dc operating point) for the circuit. In fact, PSpice always calculates the operating point because this information is needed before you proceed with other types of analysis. As we saw, the result of bias point analysis was to print values in the output file. These values fall into three categories:

- voltages at each node

- currents in each voltage source and total power dissipated

- operating points for each element

In this chapter, we explore DC Sweep analysis, which enables us to simulate a circuit for a range of dc current or voltage source values.

3.1 SWEEPING A SINGLE SOURCE

Example 2 illustrates the use of DC Sweep analysis to examine the effect of different values of a voltage source on the output voltage and current in a circuit.

EXAMPLE 2

For the circuit shown in Fig. 26, use PSpice to find the values of i_o and v_o when v_g varies from 0 to 100 V in 10 V steps.

Figure 26: The circuit for Example 2.

SOLUTION

The circuit shown in Fig. 26 is specified in schematic form in Capture, as shown in Fig. 27. Several new parts have been used in this schematic. The voltage source is modeled by the part Vsrc from the Source library, and the current source is modeled by the part Isrc from the Source library. We could also have used the parts Vdc and Idc for this example. To specify the value of the current source, highlight the part by clicking on it and select the menu option Edit/Properties. Scroll over to the

Figure 27: The circuit shown in Fig. 26 in schematic form.

Figure 28: The Property Editor with the spreadsheet of properties for the current printer part Iprint.

column labeled DC and type in the value of the current source in amps, 5. There is no need to specify the value of the voltage source, as you will see.

We have also added printers to the schematic, to cause the values of output current and output voltage to be written to the output file. The current printer has the part name Iprint, and is located in the Special library. The current printer must be placed in series with the branch through which the current of interest flows, so acts like an ammeter. It is important to orient the current printer in the direction defined for the current. Therefore, place the symbol in the schematic so that the current exits the side of the symbol with the + and − signs (see Fig. 27). To tell the printer to print the results of dc analysis, highlight the printer and select the menu option Edit/Properties. Scroll over to the column labeled DC and type a Y in this entry, as shown in Fig. 28.

There are two voltage printer parts, both in the Special library. Vprint1 prints the voltage at a single node (with respect to ground), and Vprint2 prints the voltage between two nodes. We have used Vprint2

Figure 29: The Simulation Settings dialog for the DC Sweep analysis of the schematic in Fig. 27.

in the schematic shown in Fig. 27. This part must be wired into the circuit in parallel with the component whose voltage is being measured, just like a voltmeter. Again, the orientation of this part is important, so place the symbol such that the node labeled − in the circuit schematic is wired to the side of the printer symbol with the + and − signs (see Fig. 27). You must edit the properties of the voltage printer and type a Y into the DC entry to print the voltage values in the output file.

Once the schematic is complete, create a new simulation profile for this circuit and select DC Sweep as the analysis type. Enter the name of the part whose values are to be swept, in this case the voltage source V1, and the Start Value, Final Value, and Increment, as shown in the Simulation Settings dialog in Fig. 29. Run the PSpice simulation, and look at the output file, a portion of which is shown in Fig. 30, where you can see the effect of changing the source voltage on the output voltage.

We can also depict the results of the DC Sweep analysis graphically, using Probe. In fact, after PSpice has completed the simulation, you will notice that a graph is displayed in the upper section of the simulation window, with the value of the source voltage already selected as the independent variable. Use the Trace/Add Trace menu option to specify

Figure 30: The results of DC Sweep analysis of the schematic in Fig. 27.

Figure 31: Selecting a variable to plot in Probe.

which variable to plot. As shown in Fig. 31, we selected the voltage difference across the resistor R1, using the variable name V2(R1). The

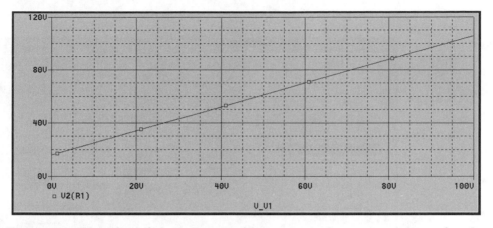

Figure 32: The plot of the output voltage versus the source voltage for the schematic in Fig. 27.

resulting plot is shown in Fig. 32.

3.2 SWEEPING MULTIPLE SOURCES

Now we alter the source file of Example 2 to sweep through values of the current and voltage sources. Example 3 presents the results.

EXAMPLE 3

The current source in the circuit shown in Fig. 26 varies from 0 to 5 A in 1 A steps. For each value of the source voltage, plot i_o as v_g varies from 0 to 100 V in steps of 20 V.

SOLUTION

The schematic is the same as the one in Fig. 27. In Capture, select the menu option PSpice/Edit Simulation Settings to specify a Secondary Sweep, the additional sweep variable and its values. The Simulation Settings dialog is shown in Fig. 33. We also changed the increment in the Primary Sweep from 10 V to 20 V. The resulting plot is shown in Fig. 34. There are five separate traces on this plot, one for each value of source current.

Figure 33: Adding a secondary sweep to the DC Sweep analysis of the schematic in Fig. 27.

Figure 34: Plot of output current as source voltage and source current are varied, for the schematic in Fig. 27.

Chapter 4

ADDITIONAL DC ANALYSIS

This chapter expands the use of dc analysis to include using PSpice to compute a Thévenin equivalent, examining the sensitivity of certain circuit output variables to changes in circuit parameter values, and calculating the effect of resistor tolerances on circuit outputs.

4.1 COMPUTING THE THÉVENIN EQUIVALENT

You may use bias analysis to calculate the open-circuit voltage and the short-circuit current at a designated pair of nodes in a circuit, and thus compute the Thévenin equivalent with respect to that pair of terminals. Example 4 illustrates this application.

EXAMPLE 4

Use PSpice to find the Thévenin equivalent with respect to terminals a, b for the circuit shown in Fig. 35.

Figure 35: The circuit for Example 5.

Figure 36: The schematic for the circuit in Fig. 35, configured to calculate the short-circuit current.

SOLUTION

A schematic created from the circuit shown in Fig. 35 is given in Fig. 36. Note that we have selected the node labeled b in Fig. 35 as the reference node. To compute the short-circuit current, we inserted a resistor between nodes a and b (node 0) whose value is much smaller than the value of other resistors in the circuit, in this case 10^{-6} Ω, effectively creating a short circuit. Bias analysis will automatically print the current through the 2 V source, which will be the short circuit current. The results of PSpice analysis, shown in Fig. 37, confirm that the short circuit current, which is the current through the voltage source V3, is 2 A.

To compute the open-circuit voltage, we cannot merely compute the voltage between nodes a and b in the circuit in Fig. 35 since PSpice requires at least two connections to every node. We have two options. First, we may connect a resistor that is much larger than the other resistors in the circuit—say, 10^6 Ω for this circuit—between nodes a and b, effectively creating an open circuit. Second, we may connect a capacitor between nodes a and b. The capacitor behaves like an open circuit during dc analysis and, therefore, does not influence the dc Thévenin

```
NODE     VOLTAGE      NODE     VOLTAGE      NODE     VOLTAGE      NODE     VOLTAGE

(N00030)   -9.8000  (N00055)   10.2000  (N00071)   -4.2000  (N00074)   -5.8000

(N00077)   30.2000  (N00091) 2.000E-06  (N00098)   -2.2000  (N00110)   -2.2000



     VOLTAGE SOURCE CURRENTS
     NAME         CURRENT

     V_V1        -3.400E+00
     V_V2        -6.000E+00
     V_V3         2.000E+00
     X_F1.VF_F1   2.000E+00

     TOTAL POWER DISSIPATION   3.04E+02   WATTS
```

Figure 37: PSpice output calculating the short circuit current.

Figure 38: The schematic for the circuit in Fig. 35, configured to calculate the open-circuit voltage.

equivalent. Here we use the first option, with the resulting schematic shown in Fig. 38. The pertinent data from the PSpice output from the simulation of the circuit in Fig. 38 is shown in Fig. 39. Note that node

```
  NODE    VOLTAGE      NODE   VOLTAGE      NODE   VOLTAGE      NODE   VOLTAGE

(N00030)   -2.0000  (N00055)   18.0000  (N00071)   -2.0000  (N00074)   38.0000

(N00077)   38.0000  (N00091)   12.0000  (N00098)-13.20E-06  (N00110)-13.20E-06



    VOLTAGE SOURCE CURRENTS
    NAME           CURRENT

    V_V1          -2.000E+00
    V_V2          -3.600E-05
    V_V3           1.200E-05
    X_F1.VF_F1     1.200E-05

    TOTAL POWER DISSIPATION   4.00E+01   WATTS
```

Figure 39: PSpice output file for determining the open-circuit voltage.

Figure 40: The Thévenin equivalent for Example 5.

a in the original circuit (Fig. 35) is assigned as node 91 by PSpice, so the open-circuit voltage is 12 V. Therefore, the Thévenin resistance is $12/2 = 6 \ \Omega$ and the Thévenin equivalent circuit is as shown in Fig. 40.

4.2   SENSITIVITY ANALYSIS

The purpose of sensitivity analysis is to obtain the dc small-signal sensitivities of each specified output variable with respect to *every* circuit parameter. Hence for circuits containing a large number of elements, tremendous amounts of output data can be generated, particularly when the sensitivity of more than one output variable is of interest. Sensitivity analysis can be performed as part of bias point analysis, and requires you to identify the voltages or currents in the circuit whose sensitivities are to be calculated. Example 5 illustrates how to perform sensitivity analysis to predict the behavior of an unloaded voltage-divider circuit.

Figure 41: The circuit for Example 5.

EXAMPLE 5

Use PSpice to study the sensitivity of the output voltage v_o in the voltage-divider circuit shown in Fig. 41.

SOLUTION

Because we have three elements in the circuit, PSpice calculates the sensitivity of v_o with respect to the independent voltage source, the 25 Ω resistor, and the 100 Ω resistor. The PSpice schematic is shown in Fig. 42. Note that we have identified and labeled the output node of

Figure 42: The PSpice schematic for the circuit in Fig. 41.

Figure 43: Specifying sensitivity analysis in PSpice.

interest for sensitivity analysis. Do this by selecting the vertical toolbar button with the double arrowhead.

To ask PSpice to perform sensitivity analysis, select Bias Point as the analysis type, and check the sensitivity analysis box, as shown in Fig. 43. We have selected V(Out) as the variable of interest. PSpice prints the data associated with sensitivity analysis in the output file The data relevant to sensitivity are shown in Fig. 44. From the simple dc analysis of the circuit, $v_o = V(Out) = 100$ V when R1, R2, and V1 are at their nominal values. From the sensitivity data, we deduce that

1. if R1 increases by 1 Ω, V(2) will decrease by 0.8 V—that is, V(2) = v_o = 99.2 V

2. if R1 increases by 1%, v_o will decrease by 0.2 V to 99.8 V

3. if R2 increases by 1 Ω, V(2) will increase by 0.2 V—that is, V(2) = v_o = 100.2 V

4. if R2 increases by 1%, v_o will increase by 0.2 V to 100.2 V

```
****      DC SENSITIVITY ANALYSIS           TEMPERATURE =    27.000 DEG C


***********************************************************************************


DC SENSITIVITIES OF OUTPUT V(OUT)

          ELEMENT          ELEMENT          ELEMENT         NORMALIZED
           NAME            VALUE          SENSITIVITY       SENSITIVITY
                                          (VOLTS/UNIT)   (VOLTS/PERCENT)

          R_R2           1.000E+02         2.000E-01         2.000E-01
          R_R1           2.500E+01        -8.000E-01        -2.000E-01
          V_V1           1.250E+02         8.000E-01         1.000E+00


      JOB CONCLUDED

      TOTAL JOB TIME            .04
```

Figure 44: Part of the output file generated by simulating the circuit shown in Fig. 42

5. if V1 increases by 1 V, V(2) will increase by 0.8 V—that is, V(2) $= v_o = 100.8$ V

6. if V1 increases by 1%, v_o will increase by 1 V to 101 V

Because we have a linear circuit, the principle of superposition applies, and therefore we may superimpose simultaneous effects. For example, let's assume that R1 increases by 1 Ω, R2 decreases by 1%, and V1 increases by 0.5 V. The effect on v_o, or V(2), will be $v_o = 100 - 0.8 - 0.2 + 0.4 = 99.4$ V.

4.3 SIMULATING RESISTOR TOLERANCES

Sensitivity analysis enabled us to examine the effect of choosing a particular component value on the output of the circuit. But we might also be interested in simulating the effect of variation in component values, called tolerances, on the output of the circuit. In the case of resistors, the PSpice model used assumes that the resistor is perfect, never varying from its desired resistance value. In the next example, we alter the model of the resistor to specify a realistic tolerance value. With this more realistic component in our schematic, we then perform worst case analysis to discover the impact of this tolerance on the output of our circuit.

Figure 45: The schematic for the circuit in Fig. 41, using a resistor model with 10% tolerance.

EXAMPLE 6

Replace the $100\,\Omega$ resistor in the circuit in Fig. 41 with a resistor having the same value but with a 10% tolerance. Use PSpice to discover the range of output voltage values that you can expect with this more realistic model of a resistor.

SOLUTION

The schematic for the circuit in Fig. 41 is shown in Fig. 45. Note that the R2 resistor displays 10% as the tolerance value. We set this tolerance value using the Property Editor to access the spreadsheet of properties for resistors. Find the column labeled Tolerance on the spreadsheet and enter 10% as its value. Then click on the Display button to cause the value of the tolerance to be displayed. The edited spreadsheet for the R2 resistor is shown in Fig. 46.

Figure 46: The edited property spreadsheet with a 10% tolerance for the R2 resistor in Fig. 45.

Next, we turn to the analysis selection. We choose DC Sweep analysis, but set the minimum and maximum values to sweep our input voltage both to 125 V, with an increment of 1 V. This screen is not shown. Then we choose the Monte Carlo/Worst Case option, as seen in Fig. 47, select the Worst Case/Sensitivity button, identify the output variable, and chose to vary devices that have device tolerances only. Once these settings have been specified, click on the More Settings button to active the dialog shown in Fig. 48 To simulate the circuit and examine the effect of resistor tolerance we perform two worst-case analyses — one to determine the maximum worst-case output voltage, and the other to determine the minimum worst-case output voltage. The parameters for the maximum worst-case analysis are shown in Fig. 48, and the relevant portion of the output file is shown in Fig. 49. This figure shows that if the value output resistor, R2, is 10% larger than its nominal value of $100\,\Omega$, the output voltage will be 101.85 V. You can confirm this result using the voltage division equation for the circuit in Fig. 41 when the R2 resistor has the value $110\,\Omega$.

Figure 47: The first step in identifying parameters for worst case analysis of resistor tolerance.

Figure 48: The second step in identifying parameters for worst case analysis of resistor tolerance.

The second worst-case analysis is specified using the parameters shown in Fig. 50. The results are shown in Fig. 51. Now we see that if the value output resistor, R2, is 10% smaller than its nominal value of $100\,\Omega$, the output voltage will be 97.826 V. You can confirm this result using the voltage division equation for the circuit in Fig. 41 when the R2 resistor has the value $90\,\Omega$. Thus, we see that when the R2 resistor in Fig. 41 has

```
****      SORTED DEVIATIONS OF V(OUT)      TEMPERATURE =   27.000 DEG C

                      WORST CASE SUMMARY

*********************************************************************


  RUN                    MAXIMUM VALUE

WORST CASE ALL DEVICES
                       101.85 at V_V1 =  125
                       ( 101.85% of Nominal)


        JOB CONCLUDED

        TOTAL JOB TIME           .07
```

Figure 49: Output of the maximum worst-case analysis of the voltage divider in Fig. 41 using a 10% output resistor.

Figure 50: Specifying a worst-case analysis which calculates the minimum output voltage value.

a tolerance of 10%, the actual output voltage may range from 97.826 V up to 101.85 V. If you were designing this circuit in an actual application, you would have to consider whether this variation around the designed value of 100 V is acceptable

```
****      SORTED DEVIATIONS OF V(OUT)      TEMPERATURE =   27.000 DEG C

                    WORST CASE SUMMARY

*********************************************************************************




   RUN                     MINIMUM VALUE

WORST CASE ALL DEVICES
                      97.826 at V_V1 =  125
                      (  97.826% of Nominal)



        JOB CONCLUDED

        TOTAL JOB TIME            .04
```

Figure 51: Output of the minimum worst-case analysis of the voltage divider in Fig. 41 using a 10% output resistor.

Chapter 5

OPERATIONAL AMPLIFIERS

PSpice offers three options for describing an operational amplifier (op amp) within a circuit schematic. The first is to model the op amp with resistors and dependent sources. The second is to take advantage of op amp models already supplied with PSpice. These models are available in a device library. The library models are considerably more sophisticated than the simple dependent-source-based models and are designed to mimic the characteristics of actual op amps. The third is to modify an existing op amp model so that it exhibits the characteristics you desire. This modified model can then be used like any other PSpice circuit element, such as a resistor, and included in multiple locations within the circuit schematic.

In the sections below, the examples illustrate each of these three options for incorporating op amps in your circuit schematics.

5.1   MODELING OP AMPS WITH RESISTORS AND DEPENDENT SOURCES

In Section 5.7 of the textbook, we introduced an equivalent circuit for the operational amplifier. That equivalent circuit is redrawn in Fig. 52.

Example 7 illustrates how to use PSpice to analyze a circuit containing an operational amplifier.

41

Figure 52: An equivalent circuit for an operational amplifier.

Figure 53: The circuit for Example 7.

EXAMPLE 7

The parameters for the operational amplifier in the circuit shown in Fig. 53 are $R_i = 200$ kΩ, $A = 10^4$, and $R_o = 5$ kΩ.

a) Use PSpice to find v_i and v_o when $v_g = 1$ V(dc).

b) Compare the PSpice solution with the analytic solution for v_i and v_o.

SOLUTION

a) The schematic for the circuit in Fig. 53 is given in Fig. 54. We used simple bias point analysis, which produced the output file, a part of which is shown in Fig. 55. Therefore,

$$V(\text{IN}) = 0.0027 \text{ V} = v_i;$$
$$V(\text{OUT}) = -9.9697 \text{ V} = v_o.$$

Figure 54: The circuit shown in Fig. 53 represented in schematic form for PSpice analysis.

```
****      SMALL SIGNAL BIAS SOLUTION      TEMPERATURE =   27.000 DEG C

*****************************************************************

NODE   VOLTAGE     NODE   VOLTAGE     NODE   VOLTAGE     NODE   VOLTAGE

(  IN)    .0027  (  OUT)   -9.9697  (N00024)   1.0000  (N00044)  -27.4180


   VOLTAGE SOURCE CURRENTS
   NAME          CURRENT

   V_V1        -9.973E-04

   TOTAL POWER DISSIPATION   9.97E-04  WATTS
```

Figure 55: The PSpice output file from the bias point analysis of the circuit in Fig. 54.

b) We obtain the analytic solution for v_i and v_o by solving the following simultaneous node-voltage equations:

$$\frac{v_i - 1}{1} + \frac{v_i}{200} + \frac{v_i - v_o}{10} = 0; \frac{v_o - v_i}{10} + \frac{v_o + 10^4 v_i}{5} + \frac{v_o}{4} = 0.$$

You should verify that the solutions for v_i and v_o are

$$v_i = 2.742 \text{ mV} \quad \text{and} \quad v_o = -9.970 \text{ V}.$$

If we assume that the operational amplifier in the circuit shown in Fig. 53 is ideal, we can alter the PSpice op amp schematic representation by making R_i and A very large and $R_o = 0$. To illustrate, we modify the schematic in Fig. 54, with R2 = 200 MΩ, A = 10^8, and R4 removed from the schematic entirely. The resulting PSpice output file after bias point analysis is shown in Fig. 56. The values of v_i and v_o from this simulation are

$$\begin{aligned} v_i &= \text{V(OUT)} = 100\text{E}^{-9} \text{ V} \approx 0 \text{ V}; \\ v_o &= \text{V(IN)} = -10.0 \text{ V}. \end{aligned}$$

These results are consistent with the analysis of the circuit shown in Fig. 53 if we assume that the operational amplifier is ideal—that is, $v_i = 0$ V and $v_o = -10$ V.

```
****        SMALL SIGNAL BIAS SOLUTION        TEMPERATURE =    27.000 DEG C

*****************************************************************************


NODE     VOLTAGE      NODE    VOLTAGE      NODE    VOLTAGE      NODE    VOLTAGE

(    IN) 100.0E-09 (   OUT)  -10.0000   (N00024)    1.0000



   VOLTAGE SOURCE CURRENTS
   NAME           CURRENT

   V_V1        -1.000E-03

   TOTAL POWER DISSIPATION    1.00E-03  WATTS
```

Figure 56: The PSpice output file for the circuit in Fig. 53, for an ideal op amp.

5.2 USING OP AMP LIBRARY MODELS

One of the most powerful features of PSpice is the library of models for various electronic devices. The full version of PSpice has models for more than 5000 devices. The evaluation version of PSpice (used to produce the examples in this supplement) has models for about 100 of the most commonly used devices. Among these models are several for op amps, including the μA741. They are all found in the Eval library. The next example repeats Example 7, but incorporates the μA741 model in the circuit schematic for Fig. 53

EXAMPLE 8

Repeat the analysis of the circuit shown in Fig. 53, but this time use the μA741 model from the PSpice model library.

SOLUTION

Figure 57 shows a schematic for the circuit in Fig. 53 that employs the PSpice part that models the μA741 op amp. As you can see, there are 7 different nodes on the op amp model. Nodes 2 and 3 correspond to the inverting and noninverting terminals of the op amp, respectively, and are wired in the schematic according to the connections to those terminals in Fig. 53. Node 6 corresponds to the output node of the op amp, and is also wired just like the corresponding node in Fig. 53. Nodes 4 and 7 are the negative and positive power supply connections, respectively. The corresponding nodes in Fig. 53, labeled −15 V and 15 V, are not explicitly wired to any other circuit components. In the schematic, these nodes must actually be connected to dc power supplies. To do this, we use node labels VCC+ and VCC-, and wire these labels to the appropriate dc power supplies to the right of the op amp circuit. This technique for drawing the schematic makes it less cluttered than it would be if we wired the dc power supplies directly into the op amp nodes. Finally, nodes 1 and 5 are not required to be connected to anything, so we leave them as is.

We can use simple bias point analysis to simulate the response of this circuit to its 1 V dc input. The relevant portion of the PSpice output file is shown in Fig. 58. Again, we discover that the output voltage is −10 V, as predicted by the analysis in the previous section.

Figure 57: The PSpice schematic for the circuit in Fig. 53, using the PSpice model of the μA741 op amp.

```
****        SMALL SIGNAL BIAS SOLUTION         TEMPERATURE =    27.000 DEG C

*****************************************************************************

NODE    VOLTAGE        NODE    VOLTAGE        NODE    VOLTAGE        NODE    VOLTAGE

(  OUT)  -10.0000   ( VCC+)   15.0000   ( VCC-)  -15.0000   (N00065)    1.0000

(N00072)-73.60E-06 (X_U1.6)     .0010  (X_U1.7)  -10.2730  (X_U1.8)  -10.2730

(X_U1.9)    0.0000 (X_U1.10)    -.6077                     (X_U1.11)  14.9600

(X_U1.12)  14.9600                     (X_U1.13)    -.5938

(X_U1.14)   -.5938                     (X_U1.53)   14.0000

(X_U1.54)  -14.0000                    (X_U1.90)   -5.4531
```

Figure 58: The PSpice output for the bias point analysis of the schematic In Flg. 57.

5.3 MODIFYING OP AMP MODELS

You can use the concept of a PSpice model to create your own model of component behavior, and associate that model with a graphical part. We illustrate this idea by modifying an existing op amp, changing its underlying circuit model and its graphical appearance. Then, in the Example 9, we use this new model to analyze the op amp circuit in Fig. 53 one more time.

Begin by placing the model of the μA741 from the Eval library in your circuit schematic. Highlight the op amp component, and select Edit/PSpice Model from the Capture menu. This will invoke the Model Editor, and you will see the description of the uA741 PSpice subcircuit, shown in Fig. 59. You define a subcircuit in a PSpice source file by first inserting a .SUBCKT control statement, which has the general form:

$$\texttt{.SUBCKT} \quad \texttt{SUBNAME} \quad \texttt{N1 N2 N3} \dots$$

where SUBNAM is the subcircuit name, and N1 N2 N3 ... are the external nodes of the subcircuit. The external nodes connect the subcircuit to the global circuit. The only restriction on selecting the external node numbers is that you cannot use zero. Any node numbers used in the description of a

```
Ex8a - PSpice Model Editor Lite - [uA741]
 File  Edit  View  Model  Plot  Tools  Window  Help

Models List
Model Name    Type
uA741*        SUBCKT

*----------------------------------------
* connections:      non-inverting input
*                   |  inverting input
*                   |  |  positive power supply
*                   |  |  |  negative power supply
*                   |  |  |  |  output
*                   |  |  |  |  |
.subckt uA741       1  2  3  4  5
*
    c1    11 12 8.661E-12
    c2     6  7 30.00E-12
    dc     5 53 dx
    de    54  5 dx
    dlp   90 91 dx
    dln   92 90 dx
    dp     4  3 dx
    egnd  99  0 poly(2) (3,0) (4,0) 0 .5 .5
    fb     7 99 poly(5) vb vc ve vlp vln 0 10.61E6 -10E6 10E6 10E6 -10E6
    ga     6  0 11 12 188.5E-6
    gcm    0  6 10 99 5.961E-9
    iee   10  4 dc 15.16E-6
    hlim  90  0 vlim 1K
    q1    11  2 13 qx
    q2    12  1 14 qx
    r2     6  9 100.0E3
    rc1    3 11 5.305E3
    rc2    3 12 5.305E3
    re1   13 10 1.836E3
    re2   14 10 1.836E3
    ree   10 99 13.19E6
    ro1    8  5 50
    ro2    7 99 100
    rp     3  4 18.16E3
    vb     9  0 dc 0

Ready                                                    NUM
```

Figure 59: The Model Editor for the uA741 op amp component.

Figure 60: The Model Editor for the newly-created IdealOpAmp subcircuit model.

subcircuit that do not appear in the `.SUBCKT` control statement are strictly local, with the exception of 0, which is always global. You terminate the subcircuit description by inserting an `.ENDS` control statement. The general form of the `.ENDS` control statement is

$$\text{.ENDS} \quad \text{<SUBNAM>}$$

where SUBNAM is optional. If included, it indicates which subcircuit is being terminated. If it is omitted, all subcircuits being described are terminated.

The subcircuit definition includes a definition of all of the components that make up the subcircuit, which in the case of the uA741 includes a large collection of components like resistors, dependent sources, and others. We wish to create a new subcircuit based on the ideal op amp, consisting of a 1 MΩ input resistor, a voltage-controlled voltage source with a gain of 10^6, and a 200 Ω output resistor. This new subcircuit, named IdealOpAmp, is shown in Fig. 60. Refer to the appendix for the netlist syntax used in

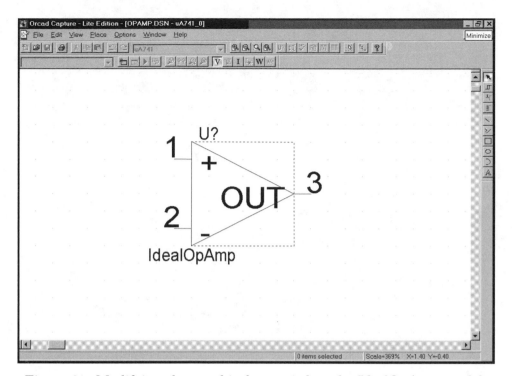

Figure 61: Modifying the graphical part tied to the IdealOpAmp model.

defining the subcircuit. Once the subcircuit definition is input, you can exit the Model Editor and return to Capture.

Next, we can edit the graphical representation of PSpice component that will be tied to the IdealOpAmp model. To do this, select Edit/Part from the Capture menu, delete the unneeded elements of the graphical representation by highlighting them and hitting the delete key, and modify the remaining elements by doubling clicking them. The result should look like the part shown in Fig. 61. Select File/Close to exit this graphical editor once you have completed the editing changes.

Finally, highlight the new IdealOpAmp component in the schematic and select Edit/Properties from the Capture menu. The last change that must be made is to modify the PSpice Template to match the subcircuit model. The uA741 subcircuit model identified the power supply terminals in the PSpice template, and those terminals are no longer needed by the model of the ideal op amp. The modified template is shown in Fig. 62.

The next example illustrates how to use the new ideal op amp component we've created.

Figure 62: The Property Editor for the IdealOpAmp component, with the PSpice Template modified to match the modified subcircuit definition.

***

EXAMPLE 9

Repeat the analysis of the circuit shown in Fig. 53, but this time use the IdealOpAmp model we have just created.

SOLUTION

With the new op amp model placed in the schematic grid, we only need to attach the remaining circuit components to the newly-created ideal op amp component. The result is shown in Fig. 63. PSpice output for the bias point analysis of the schematic in Fig. 63 is shown in Fig. 64 Again, we see that the output voltage is -10 V.

***

Figure 63: The schematic representation of the circuit in Fig. 53, using a component that represents an ideal op amp.

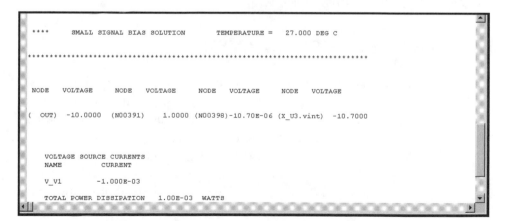

Figure 64: The PSpice output for the bias point analysis of the schematic in Fig. 63.

Chapter 6

INDUCTORS, CAPACITORS, AND NATURAL RESPONSE

Components that model inductors and capacitors are similar to those used for resistors. The primary property for these components is the value of the inductance (in henries) or the capacitance (in farads). For cases where inductors carry initial current, the initial current can be specified in the Property Editor by modifying the IC property. For cases where capacitors carry initial voltage, the Property Editor is again used to specify the initial voltage, by modifying the IC property.

6.1 TRANSIENT ANALYSIS

You use transient analysis to examine the response of a circuit as a function of time. Selecting Time Response (Transient) in the simulation settings dialog will cause PSpice to simulate the time response of the circuit. You must specify the length of time for the simulation, in seconds, and the time at which data should be stored for output. The duration of the simulation depends on the time constants in the circuit, so some preliminary analysis of the circuit is usually necessary. The start time is almost always zero, unless you want to collect data only after the circuit is in its steady-state.

You have the option to specify the step size, or time increment, to be used by PSpice, which is normally not necessary to specify. If you have used the Property Editor to set initial conditions for any inductors or capacitors in your schematic, you should check the box that will skip the initial transient analysis of the circuit. We'll see these settings in the upcoming example.

You may examine the output of a transient analysis in two different ways. First, you may direct PSpice to write the results of transient analysis to the output file, by clicking the Output File Options button in the simulation settings dialog for time response. This invokes a second dialog, in which you can specify the time increment to use when printing. Second, you may use Probe to generate plots of voltage and current versus time.

6.2 NATURAL RESPONSE

To find the natural response of a circuit, you first find the initial inductor currents and capacitor voltages. When you know these values, you can draw the schematic and edit the component properties for transient analysis. Example 10 illustrates how to use PSpice to find the natural response of a series RLC circuit. We incorporated a preliminary analytic solution as part of the example to permit checking the validity of the PSpice solution.

EXAMPLE 10

a) Analyze the natural response of the circuit shown in Fig. 65 with respect to the type of damping, the peak value of v_c, and the frequency of oscillation.

b) Use PSpice to analyze the circuit's natural response. Then use Probe to generate a plot of the voltage v_c versus t, and use the Probe cursor to identify the peak value of the voltage.

SOLUTION

a) For the series circuit shown in Fig. 65, $\alpha = R/2L = 1000$, $\alpha^2 = 10^6$, and $\omega_o^2 = 1/LC = 50 \times 10^6$. Hence the response is underdamped, and the roots of the characteristic equation are $s_1 = -1000 + j7000$ rad/s and $s_2 = -1000 - j7000$ rad/s. Therefore, the form of the solution for

Figure 65: The circuit for Example 10.

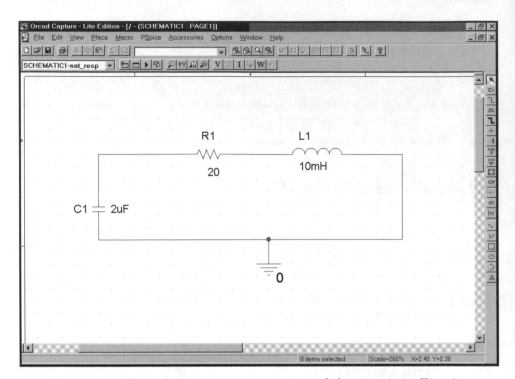

Figure 66: The schematic representation of the circuit in Fig. 65.

v_c is

$$v_c = (B_1 \cos 7000t + B_2 \sin 7000t)e^{-1000t}.$$

From the initial conditions,

$$v_c(0) = -10 \quad \text{and} \quad \frac{dv_c(0)}{dt} = 45,000 \text{ V/s}.$$

Solving for B_1 and B_2 yields

$$v_c = (-10 \cos 7000t + 5 \sin 7000t)e^{-1000t} \text{ V}, \quad t \geq 0.$$

Based on the solution for v_c, the peak value of v_c is 7.70 V, and it occurs at $t = 362.29$ μs. The damped frequency is 7000 rad/s, and thus the damped period is 897.60 μs.

b) The PSpice schematic representation of the circuit in Fig. 65 is shown in Fig. 66. For both the capacitor and the inductor, we have used the property editor to set the initial values of voltage for the capacitor and current for the inductor. To get the signs right, you must pay attention to the orientation of these parts. The initial voltage is positive at the

Figure 67: The Simulation Settings for transient analysis of the schematic in Fig. 66.

left-most terminal of the capacitor, or bottom terminal if the part is rotated once. Therefore, the initial capacitor voltage is 10 V. The initial current enters the left-most terminal of the inductor, or bottom terminal if the part is rotated once. Therefore, the initial inductor voltage is -90 mA. These values are set in the Property Editor, in the column labeled IC.

From the preliminary analysis of the circuit, we select TSTOP to be 2000 μs in the Time Response simulation settings, so that we can analyze the first two cycles of the response. We also check the box that skips the computation of the initial transient bias point. The resulting simulation setting dialog is shown in Fig. 67.

Figure 68 presents the output from the plot of capacitor voltage versus time. You invoke the cursor by selecting Trace/Cursor/Display, and then moving the cursor manually with the right and left arrow keys, or by selecting Trace/Cursor/Peak once the cursor is displayed. We see from the cursor output window that the peak output voltage is 7.7297 V, which occurs at $t = 370.286\,\mu$s, which is in good agreement with our analytic results in (a).

Figure 68: The Probe plot of the output voltage across the capacitor for the schematic in Fig. 66, with the cursor used to identify the peak value of output voltage.

Chapter 7

THE STEP RESPONSE
AND SWITCHES

The easiest way to find the step response of a circuit is to insert a dc source at the point where the step change takes place. This is illustrated in Example 11. Then, we present two of several ways to model switches in PSpice. The first involves a specially constructed voltage source, and the second involves insertion of a component that models a realistic switch. We use these two models of switches in Examples 12 and 13.

7.1 SIMPLE STEP RESPONSE

To illustrate the insertion of a dc source to model the step response, we return to the circuit used in Example 10 (Fig. 65) and define the problem in Example 11.

EXAMPLE 11

a) The switch in the circuit shown in Fig. 69 is closed at $t = 0$. At the

Figure 69: The circuit for Example 11.

instant the switch is closed, the initial current in the 10 mH inductor is 90 mA, and the voltage across the capacitor is 10 V. Figure 69 also shows the references for these initial conditions. Make a preliminary analysis of the step response, and calculate the maximum value of v_c and the time at which it occurs.

b) Create a PSpice schematic, perform transient analysis, and plot v_c versus t from 0 to 3000 μs.

c) Compare the PSpice solution with the preliminary analysis.

SOLUTION

a) From the solution of Example 10, we already know that the response is underdamped. Furthermore, we know that $s_1 = -1000 + j7000$ rad/s, that $s_2 = -1000 - j7000$ rad/s, and that the damped period is 897.60 μs. The step-response solution for v_c takes the form

$$v_c = v_f + (B'_1 \cos 7000t + B'_2 \sin 7000t)e^{-1000t}.$$

The initial values of v_c and dv_c/dt are the same as in Example 10, and the final value is 21 V. Hence $B'_1 = -31$ V and $B'_2 = 2$ V, so

$$v_c = 21 + (31 \cos 7000t - 2 \sin 7000t)e^{-1000t}, \qquad t \geq 0.$$

The maximum value of v_c is 41.22 V at 419.32 μs.

b) The PSpice schematic is shown in Fig. 70 and Figure 71 shows the plot generated by Probe. We used two cursors this time, and moved the second cursor by holding down the Shift key while using the left and right arrow keys. The cursors allow you to read the values of $v_c(\max)$ and T_d directly from the Probe plot.

c) The output of the PSpice analysis yields

$$
\begin{aligned}
v_c(\max) &= 41.361 \text{ V}, \quad \text{at } 436.78 \ \mu\text{s}; \\
T_d &= 1335.6 - 436.78 = 898.837 \ \mu\text{s}; \\
v_f &= 21 \text{ V}.
\end{aligned}
$$

These results agree with the preliminary analysis.

Figure 70: The schematic for the circuit in Fig. 69.

Figure 71: The Probe plot of the results of transient analysis on the schematic in Fig. 70.

7.2   PIECEWISE LINEAR SOURCES

The piecewise linear (PWL) time-dependent source allows you to define a single-valued function at a series of discrete times. PSpice determines the value of the source at intermediate values of time by using linear interpolation between the end points. The fact that the function must be single valued implies that the values of time used in the description of $f(t)$ must continually increase. One of the many uses for this type of source is as a model for a switch.

Example 12 illustrates the use of the piecewise linear source for modeling the switching that takes place in a transient analysis problem.

EXAMPLE 12

The circuit shown in Fig. 72 has been in operation for a long time. At $t = 0$, the 80 V source drops instantaneously to 20 V. Construct a PSpice schematic, using a PWL source, to model this circuit. The results of the analysis should be a plot of $v_o(t)$ versus t.

SOLUTION

To simulate the drop in voltage from 80 to 20 V, we insert a piecewise linear voltage source of opposite polarity in series with the 80 V dc source. We then step the piecewise source from 0 to 60 V over a short time interval. At the same time, we make the duration of the 60 V step long enough that the circuit reaches its new steady-state condition. We must now decide how rapidly the source must rise to 60 V and how long it should remain at 60 V. We leave to you verification that the time constant for the circuit shown in Fig. 72 is 40 ms. We select a rise time for the piecewise linear voltage source of 40 ms/1000, or 40 μs. The duration of the 60 V source should be much longer than 40 ms, so we arbitrarily select 10 time constants, or 400 ms. The piecewise linear voltage source has the waveform shown in Fig. 73.

Figure 72: The circuit for Example 12.

Figure 73: The waveform for the piecewise linear voltage source.

Figure 74 is the circuit schematic for the circuit shown in Fig. 72. We specify the behavior of the piecewise linear voltage source by entering pairs of time and voltage values into the appropriate columns in the Property Editor spreadsheet for this component. The first pair of values, with column headings T1 and V1, are both zero, since the value of the voltage source is 0 V at $t = 0$ s. The next pair of values, with column headings T2 and V2, are $40\,\mu$s and 60 V, since the value of the voltage source is 60 V at $t = 40\,\mu$s. The final pair of values, with column headings T3 and V3, are 40 ms and 60 V, since the voltage source continues to supply 60 V for all time after $t = 40\,\mu$s. Thus, the value for T3 is

Figure 74: The schematic for the circuit in Fig. 72.

Figure 75: The Property Editor spreadsheet for the piecewise linear voltage source from the schematic in Fig. 72.

arbitrary, so long as it is a time larger than T2. The edited spreadsheet for the piecewise linear source is shown in Fig. 75.

In this type of transient analysis problem, we use PSpice to calculate initial conditions automatically, since there were no prespecified initial conditions. Thus we don't check the SKIPBP box in the simulation settings dialog. The Probe plot in Fig. 76, which uses the cursors to determine the output voltage at $t = 40$ ms and $t = 80$ ms, yields the following results:

t, ms	$v_o(t)$, V
0	64
40	33.654
80	22.498

Furthermore, you can see from the Probe plot that $v_o(t)$ approaches 16 V as t approaches ∞.

Figure 76: Probe plot of the transient response of the schematic in Fig. 74.

7.3 REALISTIC SWITCHES

PSpice provides two special circuit components that you may use to represent actual switches more accurately. The components are both in the Eval library. One is a normally-opened switch, with the part name `Sw_tOpen` and the other is a normally-closed switch, with the part name `Sw_tClose`. These switches include attributes of a realistic switch, including the equivalent resistance of the switch when open, the equivalent resistance of the switch when closed, the time when the switch opens (if it is a normally-closed switch) or closes (if it is a normally-opened switch), and the transition time for switching from one state to the other.

We illustrate the use of the normally-opened switch in Example 13 to examine the effect of nonzero switch resistance on the step response of the *RLC* circuit.

EXAMPLE 13

Use PSpice to model the circuit shown in Fig. 69, with some modifications. Assume that the voltage source is the same as that used in Example 10, and use the same values in specifying transient analysis. But assume that the resistance of the switch in the ON position is 10 Ω. Then generate a plot of v_c versus t. Compare the results here with those in Example 10.

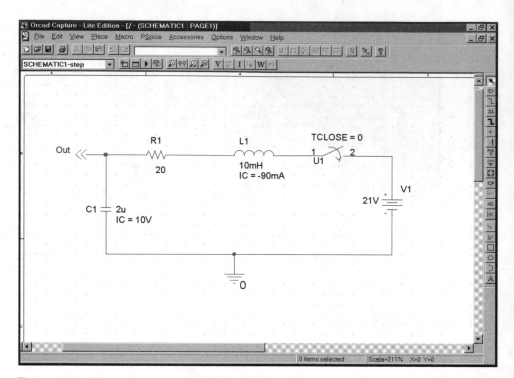

Figure 77: The schematic for the circuit in Fig. 69, using a normally-opened switch.

SOLUTION

The PSpice schematic that models the circuit in Fig. 69 is shown in Fig. 77. We set the attribute values for the switch so that when the switch is closed, its resistance is $10\,\Omega$. The spreadsheet for the switch after this attribute has been modified is shown in Fig. 78. From this figure you can see the other properties of the switch that can be specified. We have displayed the initial condition for the inductor and capacitor by highlighting the IC column in the spreadsheet and slicking the Display button.

Figure 79 shows the plot of v_c versus t. Note that the values of $v_c(\mathrm{max})$ and T_d are different in this example, as compared to Example 11, because of the additional resistance introduced by the model of the switch.

Figure 78: The property spreadsheet for the normally-opened switch in the schematic of Fig. 77.

Figure 79: The plot of capacitor voltage for the schematic in Fig. 76.

Chapter 8

VARYING COMPONENT VALUES

In PSpice, you can use a global parameter as the value of a circuit component. Then when you vary the value of the global parameter, the effect is to vary the value of the circuit component. This feature of computer-aided circuit analysis is very powerful, for it allows you to see the effects of changing a circuit parameter on the behavior of the circuit.

Example 14 demonstrates how easily PSpice can analyze the same circuit several times with a different value of a given component used in each analysis. We use Probe to illustrate graphically the varying behavior of the circuit as the component value changes.

EXAMPLE 14

Use a global parameter as the value of the resistor in Fig. 65 (see Example 10) from 20 Ω to 100 Ω in 20 Ω steps. Then plot the value of v_c versus t for each of the resistor values.

SOLUTION

The schematic for the circuit in Fig. 65 is shown in Fig. 80. While the capacitor and the inductor are modeled just as they were in Example 10, the model of the resistor is different. Instead of the resistor having a numeric value for its resistance, it has the name of a global parameter,

66

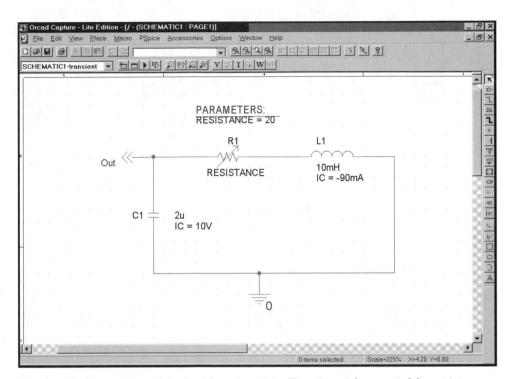

Figure 80: The schematic for the circuit in Fig. 65 with a variable resistance.

in this case RESISTANCE. We define the global parameter by adding the Param part from the Special library, as can be see in Fig. 80.

To define the global parameter, invoke the property editor on the Param component. Then, click the New button, and type the name of the global parameter in the dialog box. This creates a new property, with the name RESISTANCE, for the Param component. We then entered a default value of $20\,\Omega$ for this new property. To have this property's name and its value appear in the schematic, highlight the spreadsheet column with the new property and click the display button. Then choose Display Name and Value from the Display Format dialog. The spreadsheet for the Param part, updated as just described, is shown in Fig. 81.

Now that the schematic has completely described the circuit, we turn to specifying the analysis. We again choose Time Domain, or transient, analysis, with the same final time as we have used previously with this circuit, namely 2 ms. But now we also check the box labeled Parametric

Figure 81: The Property spreadsheet for the Param component, after the RESISTANCE property has been added.

Sweep. As shown in Fig. 82, we then select a Global parameter as the sweep variable, identify the parameter name, and input the start value, end value and increment for the sweep variable.

When we run the transient analysis, PSpice actually simulates transient analysis for five different circuits, one circuit for each of the five different resistance values we specified. We selected all five of these results to plot, as shown in Fig. 83. Note that the analysis using the smallest value of resistance has the smallest damping ratio. Increasing the value of resistance increases the damping ratio. You should confirm these results analytically.

Figure 82: Specifying a parametric sweep of the RESISTANCE global parameter for the schematic in Fig. 80.

Figure 83: Plot of capacitor voltage versus time as the resistance is varied in the series RLC circuit of Fig. 65.

Chapter 9

SINUSOIDAL STEADY-STATE ANALYSIS

We now discuss the features of PSpice that permit you to examine how a circuit behaves in response to a sinusoidal input at a fixed frequency. This behavior is the ac steady-state response of the circuit.

9.1 SINUSOIDAL SOURCES

The following equations describe the damped sinusoidal voltage source, used in simulating a circuit's time-domain response to such an input:

$$v_g = \text{VOFF}, \quad 0 \le t \le \text{TD};$$
$$v_g = \text{VOFF} + \text{VAMPL}e^{-\text{DF}(t-\text{TD})} \sin[2\pi\text{FREQ}(t-\text{TD})],$$
$$\text{TD} \le t \le \text{TSTOP};$$

where

$$\begin{aligned}
\text{VOFF} &= \text{offset voltage in volts;} \\
\text{VAMPL} &= \text{amplitude in volts;} \\
\text{FREQ} &= \text{frequency in hertz;} \\
\text{TD} &= \text{delay in seconds; and} \\
\text{DF} &= \text{damping factor in seconds}^{-1}.
\end{aligned}$$

Figure 84: A damped sinusoidal voltage.

The default values of TD and DF are zero, whereas FREQ defaults to 1/TSTOP. Figure 84 shows a graph of the damped sinusoidal voltage.

When constructing any of the time-dependent sources, you should check a plot of its waveform before using it in a specific problem. You can check the waveform by first applying it to a purely resistive circuit. The output waveform in a resistive circuit is a scaled replica of the input waveform. You may always choose a convenient scaling factor for the test circuit. For example, assume that you want to check the data statement for the damped sinusoidal voltage:

$$v_\text{g} = 12.5 + 25e^{-0.1(t-1)}\sin[0.4\pi(t-1)] \text{ V}.$$

If you apply this voltage to the voltage-divider circuit shown in Fig. 85, the output voltage is $0.8v_\text{g}$. The part name for the sinusoidal voltage source is Vsin and is found in the Source library. Again we turn to the Property Editor to access the property spreadsheet for this part and specify the values for VOFF, VAMPL, FREQ, DF, and TD. The edited spreadsheet is shown in Fig. 86. The resulting output voltage for the transient analysis of Fig. 85 is shown in Fig. 87, and confirms that we have correctly specified the properties of the sinusoidal voltage source.

9.2 SINUSOIDAL STEADY-STATE RESPONSE

The sinusoidal sources discussed in the previous section are used when a circuit's time response is of interest. Thus, these sources are intended to be used in conjunction with transient analysis. If we want to analyze a circuit to determine its sinusoidal steady-state response (also known as the ac steady-state response), we use ac voltage or current sources, whose part names are Vac and Iac, respectively. These sources are used together with AC Sweep analysis to calculate the output voltage or current phasors for one frequency

Figure 85: The schematic for a voltage divider used to check a sinusoidal source.

or over a range of different frequency values. Example 15 illustrates the use of ac sources and the AC Sweep analysis to find the single-frequency steady-state sinusoidal response.

EXAMPLE 15

The sinusoidal sources in the circuit shown in Fig. 88 are described by the equations

$$v_g = 20 \cos(10^5 t + 90°) \text{ V} \quad \text{and} \quad i_g = 6 \cos 10^5 t \text{ A}.$$

a) Use PSpice to find the magnitude and phase of v_o and i_o.

b) Check the PSpice results obtained in (a) against an analytic solution for v_o and i_o.

Figure 86: The property spreadsheet for the sinusoidal voltage source in Fig. 85. (Right-click in the spreadsheet and select Pivot to see this display format.)

Figure 87: The plot of the output voltage for the transient analysis of the schematic in Fig. 85.

Figure 88: The circuit for Example 15.

SOLUTION

a) The schematic representation of the circuit in Fig. 88 is shown in Fig. 89. We edit the property spreadsheets for the ac sources to specify the magnitude and phase angle. Note that the phase angle property assumes that the source is specified as a cosine, and the phase angle units are degrees. An edited spreadsheet for the ac voltage source is shown in Fig. 90. Note further from the schematic in Fig. 89 that we have inserted current and voltage printers to cause the results of the

Figure 89. The schematic for the circuit in Fig. 88.

Figure 90: The edited property spreadsheet for the ac voltage source in the schematic of Fig. 89.

analysis to be printed in the output file. In order to get the desired output data, we must edit the property spreadsheets for these parts. Remember to specify the output from AC analysis by placing a "Y" (for "yes") in the column labeled AC. Place a Y in the columns labeled MAG and PHASE if you want the output values in polar form. Place a Y in the columns labeled REAL and IMAG if you want the output values in rectangular form. The edited property spreadsheet for the voltage printer is shown in Fig. 91. As you can see, we have specified that both the polar and rectangular forms of the output voltage be written to the output file.

Next we specify the type of analysis, which in this case is AC Sweep. The dialog box for this analysis is shown in Fig. 92. We are asked to specify the start and end frequency and the total number of frequency points in our sweep. Since we want PSpice to analyze the schematic for only a single frequency, the start and end frequencies are the same, and the total number of frequency values is 1. Note that we must specify the frequency in a PSpice source file in hertz. The pertinent output from the

Figure 91: The edited property spreadsheet for the voltage printer in the schematic of Fig. 89.

PSpice analysis is shown in Fig. 93. The PSpice simulation results are $\mathbf{V}_\text{o} = 16.31\underline{/71.51°}$ V and $\mathbf{I}_\text{o} = 3.261\underline{/124.6°}$ A. b) We obtain the solution for the phasor voltage $\mathbf{V}_\text{o}$ by solving a single node-voltage equation:

$$\frac{\mathbf{V}_\text{o}}{4 + j3} + \frac{\mathbf{V}_\text{o}}{3 - j4} + \frac{\mathbf{V}_\text{o} - j20}{1} = 6\underline{/0°}.$$

The solution for $\mathbf{V}_\text{o}$ is

$$\mathbf{V}_\text{o} = 16.31\underline{/71.51°} \text{ V}.$$

Hence

$$\mathbf{I}_\text{o} = \frac{\mathbf{V}_\text{o}}{3 - j4} = 3.26\underline{/124.64°} \text{ A}.$$

Figure 92: The dialog box for AC Sweep analysis of the circuit in Fig. 89.

Figure 93: The output from the ac analysis of the schematic in Fig. 90.

Chapter 10

COMPUTING AC POWER WITH PROBE

The interactive graphics capabilities provided by Probe offer the opportunity to study the relationships among the various ac power quantities. In this chapter, we investigate the relationships among current, voltage, instantaneous real power, average power, instantaneous reactive power, and reactive power for the circuit shown in Fig. 94. Although we introduce several new features of Probe, you do not need any new PSpice constructs in order to investigate ac power.

To begin, construct a schematic for the parallel RC circuit in Fig. 94, as shown in Fig. 95. Specify transient analysis of this circuit for 100 μs.

When the blank graph appears with the main menu below it, select the menu item Plot/Add Plot to Window three times to create four blank plots on the screen. You can plot many different variables in these four plots and compare them. First, select the top plot on the screen, and plot the voltage at node 1. The plot should look like the first one (top) shown in Fig. 96. Next, select the second plot on the screen, and plot the current through the source. Note the phase relationship between the source voltage and current: The current leads the voltage because the load is capacitive. Compute the

Figure 94: A circuit used to investigate AC power relationships.

Figure 95: The schematic for the circuit in Fig. 94.

power factor angle of the load as follows:

$$
\begin{aligned}
Z_{\mathrm{L}} &= R \| 1/j\omega C \\
&= \frac{R}{j\omega RC + 1}; \\
\underline{/Z_{\mathrm{L}}} &= 0 - \arctan(\omega RC) \\
&= -\arctan(10^5 \cdot 40.5 \cdot 0.185 \times 10^{-6}) \\
&= -36.87^\circ.
\end{aligned}
$$

Now, select the third plot, and graph the instantaneous power supplied by the source. Recall that the expression for instantaneous power is $v(t)\cdot i(t)$. Probe allows you to supply algebraic expressions when it prompts for the variable to plot, so you simply type -V2(R1)*I(V1). Note that the minus sign (-) inverts the sign of the instantaneous power to conform to the PSpice sign convention. In this plot, you have confirmed that the frequency of the instantaneous power is double the frequency of the voltage and current.

Figure 96: Some ac power characteristics from the circuit shown in Fig. 94: top plot, source voltage; second plot, source current; third plot, instantaneous and average power supplied by the source; bottom plot, instantaneous and reactive power absorbed by the capacitor.

Now display the average power supplied by the source on top of the plot of the instantaneous power. The formula for average power is

$$P = \frac{1}{2} V_{\mathrm{m}} I_{\mathrm{m}} \cos(\underline{/Z_{\mathrm{L}}}).$$

Select the top two plots, one at a time, and use the cursor to find the maximum values $V_{\mathrm{m}} = 180$ and $I_{\mathrm{m}} = 5.56$. Select the third plot again, and add the trace described by the expression

```
.5*180*5.56*cos(-36.87/57.3)
```

Note that we divided the impedance angle ($\underline{/Z_{\mathrm{L}}} = 36.87°$) by 57.3 to convert the angle from degrees to radians, as required by Probe. In the third plot shown in Fig. 96, the equation for average power produced a straight line that runs through the center of the plot of instantaneous power, as you might have expected. You can use the cursor to find the value of this average power is 408 V.

Use the fourth plot to examine the reactive power in the circuit. On the fourth plot, generate a plot of the instantaneous power absorbed by the capacitor. Note that this is instantaneous *reactive* power, whose average value is zero. On top of this plot, place the plot of the reactive power supplied by the source, which is described by the expression

```
.5*180*5.56*sin(-36.87/57.3)
```

You have completed the plot shown in Fig. 96. Now try to describe the relationship between the instantaneous and average reactive power absorbed by the capacitor.

This exploration of the ac power characteristics of the circuit shown in Fig. 96 using the power of Probe necessarily has been brief. You should take the time to investigate further; try to construct plots of complex power, for example.

Chapter 11

SUMMARY

OrCad PSpice A/D is one of numerous software packages designed to simulate an electric circuit or network. One of the helpful features of these simulation programs is that they allow the user to define the circuit in terms of a topological, as opposed to a mathematical, description. Although the primary purpose of PSpice is to facilitate the analysis of circuits containing nonlinear electronic components, we have introduced it to you as a tool for analyzing linear, lumped-parameter circuits. There are several reasons for this early introduction. First, it acquaints you with a powerful computational tool. Second, you can test the computer-generated values by direct computation, thus gaining confidence in your ability to formulate correct source files. Third, the computer makes it possible for you to attack circuit problems that are computationally very complex and lengthy if attempted without the aid of a computer. These complex problems often add significantly to your understanding of circuit behavior. Finally, simulation software is part of modern engineering practice. The introduction of such software at an elementary level is a first step in preparing you for what lies ahead.

BIBLIOGRAPHY

OrCAD PSpice A/D User's Guide, Release 9, Beaverton, Oregon, 1998.

APPENDIX

QUICK REFERENCE TO PSPICE NETLIST STATEMENTS

Before PSpice can begin analysis, the schematic must be translated into a collection of statements which identify the types of components, their attributes, and their interconnections. This collection is the netlist. This appendix presents the netlist syntax for most of the schematic parts presented in this supplement. Being familiar with the netlist syntax is helpful when debugging errors in your schematic, and is necessary when creating new subcircuit models or modifying existing subcircuit models for parts.

DATA STATEMENTS

In the syntax for data statements, *xxx* represents any alphanumeric string used to identify uniquely a circuit component, n+ represents the positive circuit node, and n− represents the negative voltage node. Positive voltage drop is from n+ to n−, and positive current flows from n+ to n−. Any elements in the syntax specification that are optional are enclosed by brackets ([]).

INDEPENDENT DC VOLTAGE AND CURRENT SOURCES

SYNTAX (VOLTAGE) Vxxx   n+   n−   DC   value

SYNTAX (CURRENT) Ixxx   n+   n−   DC   value

    where

        value is the dc voltage in volts for the voltage source and the dc current in amperes for the current source.

DESCRIPTION Provides a constant source of voltage or current to the circuit.

EXAMPLES V_in $N_0001 $N_0002   DC   3.5

         I_source $N_0003   0   DC   0.25

INDEPENDENT AC VOLTAGE AND CURRENT SOURCES

SYNTAX (VOLTAGE) Vxxx   n+   n−   AC   mag   [phase]

SYNTAX (CURRENT) Ixxx   n+   n−   AC   mag   [phase]

> where

> mag is the magnitude of the ac waveform in volts for the voltage source and in amperes for the current source;

> phase is the phase angle of the ac waveform in degrees and has a default value of 0.

DESCRIPTION Supplies a sinusoidal voltage or current at a fixed frequency to the circuit.

EXAMPLES V_ab $N_0003 $N_0002   AC   10   90

        I_12 $N_0004 $N_0006   AC   0.5

INDEPENDENT TRANSIENT VOLTAGE AND CURRENT SOURCES

SYNTAX (VOLTAGE) Vxxx   n+   n−   trans_type

SYNTAX (CURRENT) Ixxx   n+   n−   trans_type

> where trans_type is one of the following transient waveform types:

> - *Exponential* **EXP** (start peak [delay1] [rise] [delay2] [fall])
>   where
>   start is the initial value of voltage in volts or current in amperes;
>   peak is the maximum value of voltage in volts or current in amperes;
>   delay1 is the delay in seconds prior to a change in voltage or current from the initial value, which is 0 seconds by default;

rise is the time constant in seconds of the exponential decay from start to peak, which has a default value of tstep seconds (see .TRAN);

delay2 is the delay in seconds prior to introducing a second exponential decay, from the maximum value toward the initial value, which has a default value of delay1 + tstep seconds (see .TRAN);

fall is the time constant in seconds of the exponential decay from peak to final, which has a default value of tstep seconds (see .TRAN).

- *Pulsed* PULSE(min max [delay] [rise] [fall] [width] [period])
  where

  min is the minimum value of the waveform in volts for the voltage source and in amperes for the current source;

  max is the maximum value of the waveform in volts for the voltage source and in amperes for the current source;

  delay is the time in seconds prior to the onset of the pulse train, which has a default of 0 seconds;

  rise is the time in seconds for the waveform to transition from min to max, which has a default value of tstep seconds (see .TRAN);

  fall is the time in seconds for the waveform to transition from max to min, which has a default value of tstep seconds (see .TRAN);

  width is the time in seconds that the waveform remains at the max value, which has a default value of tstop seconds (see .TRAN);

  period is the time in seconds that separates the pulses in the pulse train, which has a default of tstop seconds (see .TRAN).

- *Piecewise linear* PWL(t1 val1 t2 val2 . . .)
  where

  t1 val1 are a paired time (in seconds) and value (in volts for a voltage source and amperes for a current source) that specifies a corner of the waveform; all pairs of times and values are linearly connected to form the whole waveform.

- *Damped sinusoid* SIN(off peak [freq] [delay] [damp] [phase])
  where

off is the initial value of the voltage in volts or the current in amperes;

peak is the maximum amplitude of the voltage in volts or the current in amperes;

freq is the sinusoidal frequency in hertz, which has a default value of 1/tstop hertz (see .TRAN);

delay is the time in seconds that the waveform stays at the initial value before the sinusoidal oscillation begins, which has a default value of 0 seconds;

damp is the sinusoidal damping factor in seconds^{-1} used to specify the decaying exponential envelope for the sinusoid, which has a default value of 0 seconds^{-1};

phase is the initial phase angle of the sinusoidal waveform in degrees, which has a default value of 0 degrees.

DESCRIPTION Supplies a time-varying voltage or current to the circuit whose waveform can be characterized as exponential, pulsed, piecewise linear, or damped sinusoidal.

EXAMPLES

```
V_s    $N_0003  $N_0001  EXP(2  6  .5  .1  .5  .2)

I_in   $N_0004  0  PULSE(-.3  .3  0  .01  .01  1  2)

I_5    $N_0002  $N_0003  PWL(0 .2 1 .6 1.5 6 3 -.5)

V_g    0    $N_0003 SIN(0  2  100  0  0  90)
```

DEPENDENT VOLTAGE-CONTROLLED VOLTAGE SOURCE

SYNTAX Exxx   n+   n−   cn+   cn−   gain

where

cn+ is the positive node for the controlling voltage;

cn− is the negative node for the controlling voltage;

gain is the ratio of the source voltage (between n+ and n−) to the controlling voltage (between cn+ and cn−).

DESCRIPTION Provides a voltage source whose value depends on a voltage measured elsewhere in the circuit.

EXAMPLE E_op $N_0001 $N_0002 $N_0004 0 .5

DEPENDENT CURRENT-CONTROLLED CURRENT SOURCE

SYNTAX Fxxx n+ n− Vyyy gain

> where
>
> Vyyy is the name of the voltage source through which the controlling current flows;
>
> gain is the ratio of the source current (flowing from n+ to n−) to the controlling current (flowing through Vyyy).

DESCRIPTION Provides a current source whose value depends on the magnitude of a current flowing through a voltage source elsewhere in the circuit.

EXAMPLE F_dep $N_0003 $N_0002 V_control 10

DEPENDENT VOLTAGE-CONTROLLED CURRENT SOURCE

SYNTAX Gxxx n+ n− cn+ cn− gain

> where
>
> cn+ is the positive node for the controlling voltage;
>
> cn− is the negative node for the controlling voltage;
>
> gain is the ratio of the source current (flowing from n+ to n−) to the controlling voltage (between cn+ and cn−), in ohms^{-1}.

DESCRIPTION Provides a current source whose value depends on the magnitude of a voltage measured elsewhere in the circuit.

EXAMPLE G_on $N_0003 $N_0006 $N_0004 $N_0001 0.35

DEPENDENT CURRENT-CONTROLLED VOLTAGE SOURCE

SYNTAX Hxxx   n+   n−   Vyyy   gain

> where
>
> Vyyy is the name of the voltage source through which the controlling current flows;
>
> gain is the ratio of the source voltage (between n+ and n−) to the controlling current (flowing through Vyyy), in ohms.

DESCRIPTION Provides a voltage source whose value depends on the magnitude of current flowing through another voltage source elsewhere in the circuit.

EXAMPLE H_out $N_0007 $N_0002   V_dummy   −2.5E-3

RESISTOR

SYNTAX Rxxx   n+   n−   [mname]   value

> where
>
> mname is the name of a **RES** model defined in a .**MODEL** statement (see .**MODEL**)—note that mname is optional;
>
> value is the resistance, in ohms.

DESCRIPTION Models a resistor, a circuit element whose voltage and current are linearly dependent.

EXAMPLE R_for $N_0003 $N_0004   16E3

INDUCTOR

SYNTAX Lxxx   n+   n−   [mname]   value   [IC = icval]

> where
>
> mname is the name of an **IND** model defined in a .**MODEL** statement (see .**MODEL**)—note that mname is optional;
>
> value is the inductance, in henries;
>
> icval is the initial value of the current in the inductor, in amperes—note that specifying the initial condition is optional.

DESCRIPTION Models an inductor, a circuit element whose voltage is linearly dependent on the derivative of its current.

EXAMPLE L_44 $N_0001 $N_0009   3E-3   IC = 1E-2

CAPACITOR

SYNTAX Cxxx   n+   n−   [mname]   value   [IC = icval]

where

mname is the name of a CAP model defined in a .MODEL statement (see .MODEL)—note that mname is optional;

value is the capacitance, in farads;

icval is the initial value of the voltage across the capacitor, in volts— note that specifying the initial condition is optional.

DESCRIPTION Models a capacitor, a circuit element whose current is linearly dependent on the derivative of its voltage.

EXAMPLES C_two $N_0004   0   cmod   2E-6

.MODEL   cmod   CAP(C = 1)

MUTUAL INDUCTANCE

SYNTAX Kxxx   Lyyy   Lzzz   value

where

Lyyy is the name of the inductor on the primary side of the coil;

Lzzz is the name of the inductor on the secondary side of the coil;

value is the mutual coupling coefficient, k, which has a value such that $0 \leq k < 1$.

DESCRIPTION Models the magnetic coupling between any two inductor coils in a circuit.

EXAMPLE K_ab   L_a   L_b   0.5

SUBCIRCUIT DEFINITION

SYNTAX `.SUBCKT`   name   [nodes]   `.ENDS`

> where
>
> name is the name of the subcircuit, as referenced by an **X** statement (see subcircuit call);
>
> nodes is the optional list of nodes used to identify the connections to the subcircuit;
>
> `.ENDS` signifies the end of the subcircuit definition.

DESCRIPTION Used to provide "subroutine"-type definitions of portions of a circuit. When the subcircuit is referenced in an **X** statement, the definition between the `.SUBCKT` statement and the `.ENDS` statement replaces the **X** statement in the source file.

EXAMPLES

`.SUBCKT   opamp   $N_0001   $N_0002   $N_0003   $N_0004   $N_0005`

`.ENDS`

SUBCIRCUIT CALL

SYNTAX Xxxx   nodes   name

> where
>
> nodes is the optional list of circuit nodes used to connect the subcircuit into the rest of the circuit; there must be as many nodes in this list as there are in the subcircuit definition (see "Subcircuit Definition");
>
> name is the name of the subcircuit as defined in a `.SUBCKT` statement.

DESCRIPTION Replaces the **X** statement with the definition of a subcircuit, which permits a subcircuit to be defined once and used many times within a given source file.

EXAMPLE

`X_amp   $N_0004   $N_0002   $N_0008   $N_0005   $N_0001   opamp`

LIBRARY FILE

SYNTAX .LIB   [fname]

> where
>
>> fname is the name of the library file containing .MODEL or .SUBCKT statements referenced in the source file, which by default is the nominal or evaluation library file.

DESCRIPTION Used to reference models or subcircuits.

EXAMPLE .LIB   mylib.lib

DEVICE MODELS

SYNTAX .MODEL   mname   mtype   [(par=value)]

> where
>
>> mname is a unique model name, which is also used in the device statement that incorporates this model;
>>
>> mtype is one of the model types available;
>>
>> par=value is an optionally specified list of parameters and their assigned values, specific to the model type:
>>
>>> - RES, which models a resistor and has the parameter R, the resistance multiplier, whose default value is 1;
>>> - IND, which models an inductor and has the parameter L, the inductance multiplier, whose default value is 1;
>>> - CAP, which models a capacitor and has the parameter C, the capacitance multiplier, whose default value is 1.

DESCRIPTION Defines standard devices that can be used in a circuit and sets parameter values that characterize the specific device being modeled.

EXAMPLES .MODEL   lmodel   IND(L = 2)

> .MODEL   resist   RES

CONTROL STATEMENTS

DC ANALYSIS

SYNTAX .DC   [type]   vname   start   end   incre   [nest-sweep]

> where
>
>> type is the sweep type and must be LIN for a linear sweep from the start value to the end value, OCT for a logarithmic sweep from start value to end value in octaves, DEC for a logarithmic sweep from start value to end value in decades, or LIST if a list of values is to be used; the default sweep type is LIN;
>>
>> vname is the name of the circuit element whose value is swept, usually an independent voltage or current source;
>>
>> start is the beginning value of vname;
>>
>> end is the final value of vname; note that when the sweep type is LIST, the start, end, and incre values are replaced by a list of values that vname will take on;
>>
>> incre is the step size for the LIN sweep type and is the number of points per octave or decade for the OCT and DEC sweep types;
>>
>> nest-sweep is an optional additional sweep of a second circuit variable, which follows the same syntax as the sweep for the first variable.

DESCRIPTION Provides a method for varying one or two dc source values and analyzing the circuit for each sweep value.

EXAMPLES .DC   V_ab $N_0001 $N_0010   .5

.DC   I_1   DEC   .1   10   30   V_1   DEC   10   100   30

SENSITIVITY

SYNTAX .SENS   vname

> where vname is the circuit variable name (or list of circuit variable names) for which sensitivity analysis will be performed.

DESCRIPTION Computes and prints to the output file a dc sensitivity analysis of vname to the values of other circuit elements, including resistors, independent voltage and current sources, and voltage- and current-controlled switches.

EXAMPLE .SENS   V($N_0002)   I(V_in)

TRANSIENT ANALYSIS

SYNTAX .TRAN   tstep   tstop   [npval]   istep   [UIC]

where

tstep is the time interval separating values that are printed or plotted;

tstop is the ending time for the transient analysis;

npval is the time between 0 and the first value printed or plotted, which by default is 0;

istep is the internal time step used for computing values, which by default is tstop/50;

UIC will bypass the calculation of the bias point, which usually precedes the transient analysis, and use the initial conditions specified by IC = in inductor and capacitor data statements instead.

DESCRIPTION Performs a transient analysis of the circuit described in the source file to calculate the values of circuit variables as a function of time.

EXAMPLE .TRAN   .01   10   0   .001   UIC

INITIAL CONDITION

SYNTAX .IC   Vnode=value

where

Vnode=value is one or a list of pairs consisting of a voltage node Vnode, represented in standard form, and the initial value in volts at that node.

DESCRIPTION Sets the initial conditions for transient analysis by specifying one or more node voltage values.

EXAMPLE .IC V($N_0001) = 10

AC ANALYSIS

SYNTAX .AC   [type]   num   start   end

where

type is the type of sweep, which must be one of the following keywords:

LIN for a linear sweep in frequency, which is the default;

OCT for a logarithmic sweep in frequency by octaves;

DEC for a logarithmic sweep in frequency by decades;

num is the total number of points in the sweep for a linear sweep and the number of points per octave or decade for a logarithmic sweep;

start is the starting frequency, in hertz;

end is the final frequency, in hertz.

DESCRIPTION Computes the frequency response of the circuit described in the source file as the frequency is swept either linearly or logarithmically from an initial value to a final value.

EXAMPLE .AC   LIN   300   10   1000

STEPPED VALUES

SYNTAX .STEP   [type]   name   start   end   incre

where

type is one of the following keywords describing the type of parameter variation:

LIN for a linear variation, which is the default;

OCT for a logarithmic variation by octaves;

DEC for a logarithmic variation by decades;

name is the name of the circuit element whose value is to be varied;

start is the starting value for the circuit element, in units appropriate to the type of circuit element;

end is the final value for the circuit element, in units appropriate to the type of circuit element;

incre is the step size for a linear variation and the number of values per octave or decade for the logarithmic variation. Note that when a discrete list of parameter values is desired, type is not used, and the name is followed by the keyword LIST and a list of parameter values.

DESCRIPTION Used to step a circuit element's value through a range either linearly or logarithmically or through a discrete list, analyzing the circuit for each value.

EXAMPLES `.STEP  DEC  RES  RMOD(R)  10  10E4  3`

`.STEP  V_2  LIST  1  4  12  19`

FOURIER SERIES ANALYSIS

SYNTAX `.FOUR`   freq   vname

> where
>
> freq is the fundamental frequency of the circuit;
>
> vname is the variable name or the list of variable names for which Fourier series coefficients will be computed.

DESCRIPTION Uses the results of a transient analysis to compute the Fourier coefficients for the first nine harmonics. Note the `.FOUR` statement requires a `.TRAN` statement to perform the transient analysis.

EXAMPLE `.FOUR  10E3  V($N_0003)`

OUTPUT STATEMENTS

OPERATING POINT

SYNTAX `.OP`

DESCRIPTION Outputs information describing the bias point computations for the circuit being simulated.

EXAMPLE `.OP`

PRINT RESULTS

SYNTAX `.PRINT`   type   vname

> where
>
> type identifies the type of analysis performed to generate the data being printed and must be one of the keywords `DC`, `AC`, or `TRAN`;
>
> vname is the variable name or the list of variable names for which values are to be printed.

DESCRIPTION Prints the results of circuit analysis in table form to an output file for each program variable specified.

EXAMPLE `.PRINT    AC    V(R_in)  I(V_source)`

PROBE

SYNTAX .PROBE   [vname]

> where vname is the optionally specified variable or list of variables
> whose values from dc, ac, and transient analysis will be stored in
> the file that PROBE uses to generate plots. If no variable name
> is included, values of all circuit variables will be stored in the
> PROBE file.

DESCRIPTION Generates a file of data from dc, ac, and transient analysis used by PROBE to generate high-quality plots.

EXAMPLE .PROBE V($N_0002) I(R_in) I(R_out)

MISCELLANEOUS STATEMENTS

TITLE LINE

DESCRIPTION Each source file must have a title line as its first line.

EXAMPLE A Circuit to Simulate RLC Frequency Response

END STATEMENT

SYNTAX .END

DESCRIPTION Each source file must have an .END statement as its final
line.

EXAMPLE .END

Index

Warning and Disclaimer

This book is designed to provide tutorial information about the Cadence™ set of computer programs. Every effort has been made to make this book as complete and accurate as possible, but no warranty nor fitness is implied.

The information is provided on an "as is" basis. The authors and Prentice Hall shall have neither the liability nor responsibility to any person or entity with respect to any loss or damages arising from the information contained in this book or from the use of the disks and programs that may accompany it.

LIMITED WARRANTY AND DISCLAIMER OF WARRANTY: Because this SOFTWARE is being given to you without charge, the Company makes no warranties about the SOFTWARE, which is provided "AS-IS." THE COMPANY DISCLAIMS ALL WARRANTIES OR MERCHANTABILITY AND FITNESS FOR A PARTICULAR PURPOSE. THE COMPANY DOES NOT WARRANT, GUARANTEE, OR MAKE ANY REPRESENTATION REGARDING THE USE OF OR RESULT OF THE USE OF THE SOFTWARE. IN NO EVENT SHALL THE COMPANY OR ITS EMPLOYEES, AGENTS, SUPPLIERS, OR CONTRACTORS BE LIABLE FOR ANY INCIDENTAL, INDIRECT, SPECIAL, OR CONSEQUENTIAL DAMAGES ARISING OUT OF OR IN CONNECTION WITH THE LICENSE GRANTED UNDER THIS AGREEMENT INCLUDING, WITHOUT LIMITATION, LOSS OF USE, LOSS OF DATA, LOSS OF INCOME OR PROFIT, OR OTHER LOSSES SUSTAINED AS A RESULT OF INJURY TO ANY PERSON, COMPANY, OR AN AUTHORIZED REPRESENTATIVE OF THE COMPANY HAD BEEN ADVISED OF THE POSSIBILITY OF SUCH DAMAGES.

SOME JURISDICTIONS DO NOT ALLOW THE LIMITATION OF IMPLIED WARRANTIES OR LIABILITY FOR INCIDENTAL, INDIRECT, SPECIAL, OR CONSEQUENTIAL DAMAGES, SO THE ABOVE LIMITATIONS MAY NOT ALWAYS APPLY. THE WARRANTIES IN THIS AGREEMENT GIVE YOU SPECIFIC LEGAL RIGHTS AND YOU MAY ALSO HAVE OTHER RIGHTS, WHICH VARY IN ACCORDANCE WITH LOCAL LAW.

Customer Support

Please visit www.orcad.com/technical/technical.asp or www.pspice.com/online_help for more information and advice about Orcad and PSpice products.

If you are a student using our products, you may submit questions via email to tech.support@pspice.com.

Trademarks

"Adobe" and "Acrobat" are registered trademarks of Adobe Systems, Incorporated.

"MicroSim PSpice," "Orcad," "Orcad Layout," "Orcad Express," "Orcad Capture," "Orcad PSpice," and "Orcad PSpice A/D" are registered trademarks and "Designlab," "Orcad Capture CIS," and "Orcad Express CIS" are trademarks of Orcad, Inc.

"Probe," "Schematics," "StmEd," "Stimulus Editor," "Parts," and "Monte Carlo" are trademarks of the MicroSim Corporation.

"Microsoft," "MS," and "MS-DOS" are registered trademarks and "Windows" and "Win32s" are trademarks of the Microsoft Corporation.

"IBM" is a registered trademark of the International Business Machines Corporation.

"Sun" is a registered trademark and "Open Windows" is a trademark of Sun Microsystems Incorporated.

LICENSE AGREEMENT AND LIMITED WARRANTY

READ THIS LICENSE CAREFULLY BEFORE OPENING THE CD SOFTWARE PACKAGE. BY OPENING THE CD SOFTWARE PACKAGE, YOU ARE AGREEING TO THE TERMS AND CONDITIONS OF THIS LICENSE. IF YOU DO NOT AGREE, DO NOT OPEN THE CD SOFTWARE PACKAGE. THESE TERMS APPLY TO ALL LICENSED SOFTWARE ON THE DISC EXCEPT THAT THE TERMS FOR USE OF ANY SHAREWARE OR FREEWARE ON THE CD ARE AS SET FORTH IN THE RELEVANT LICENSE LOCATED IN THIS APPENDIX AND/OR IN THE RELEVANT ELECTRONIC LICENSE LOCATED ON THE CD ("FREEWARE LICENSES"):

1. **Grant of License and Ownership**: The enclosed computer programs ("Software") are licensed, not sold, to you by Prentice-Hall, Inc. ("We" or the "Company") in consideration of your purchase of this book, and your agreement to these terms. You own only the disc but we and/or our licensors own the Software itself. This license grants to you a nonexclusive license to use and display the enclosed copy of the Software on a single computer (i.e., with a single CPU), for educational use only, so long as you comply with the terms of this Agreement. You may make one copy for back up only. We reserve any rights not granted to you. The applicable Freeware License may grant you additional rights as between you and the shareware authors.

2. **USE RESTRICTIONS**: You may not sell or license copies of the Software or the Documentation to others, except as provided in any applicable Freeware License. You may not transfer, distribute or make available the Software or the Documentation, except as provided in any applicable Freeware License. You may not reverse engineer, disassemble, decompile, modify, adapt, translate, or create derivative works based on the Software or the Documentation, except as provided in any applicable Freeware License. You may be held legally responsible for any copying or copyright infringement which is caused by your failure to abide by the terms of these restrictions.

3. **TERMINATION**: This license is effective until terminated. This license will terminate automatically without notice from the Company if you fail to comply with any provisions or limitations of this license. Upon termination, you shall destroy the Documentation and all copies of the Software. All provisions of this Agreement as to limitation and disclaimer of warranties, limitation of liability, remedies or damages, and our ownership rights shall survive termination.

4. **DISCLAIMER OF WARRANTY: THE COMPANY AND ITS LICENSORS MAKE NO WARRANTY ABOUT THE SOFTWARE, WHICH IS PROVIDED "AS-IS." IF THE DISC IS DEFECTIVE IN MATERIALS OR WORK-MANSHIP, YOUR ONLY REMEDY IS TO RETURN IT TO THE COMPANY WITHIN 30 DAYS FOR REPLACEMENT UNLESS THE COMPANY DETERMINES IN GOOD FAITH THAT THE DISC HAS BEEN MISUSED OR IMPROPERLY INSTALLED, REPAIRED, ALTERED, OR DAMAGED. THE COMPANY DISCLAIMS ALL WARRANTIES, EXPRESS OR IMPLIED, INCLUDING WITHOUT LIMITATION, THE IMPLIED WARRANTIES OF MERCHANTABILITY AND FITNESS FOR A PARTICULAR PURPOSE. THE COMPANY DOES NOT WARRANT, GUARANTEE OR MAKE ANY REPRESENTATION REGARDING THE ACCURACY, RELIABILITY, CURRENTNESS, USE, OR RESULTS OF USE, OF THE SOFTWARE.**

5. **LIMITATION OF REMEDIES AND DAMAGES: IN NO EVENT, SHALL THE COMPANY OR ITS EMPLOYEES, AGENTS, LICENSORS, OR CONTRACTORS BE LIABLE FOR ANY INCIDENTAL, INDIRECT, SPECIAL, OR CONSEQUENTIAL DAMAGES ARISING OUT OF OR IN CONNECTION WITH THIS LICENSE OR THE SOFTWARE, INCLUDING, WITHOUT LIMITATION, LOSS OF USE, LOSS OF DATA, LOSS OF INCOME OR PROFIT, OR OTHER LOSSES SUSTAINED AS A RESULT OF INJURY TO ANY PERSON, OR LOSS OF OR DAMAGE TO PROPERTY, OR CLAIMS OF THIRD PARTIES, EVEN IF THE COMPANY OR AN AUTHORIZED REPRESENTATIVE OF THE COMPANY HAS BEEN ADVISED OF THE POSSIBILITY OF SUCH DAMAGES.** SOME JURISDICTIONS DO NOT ALLOW THE LIMITATION OF DAMAGES IN CERTAIN CIRCUMSTANCES, SO THE ABOVE LIMITATIONS MAY NOT ALWAYS APPLY.

6. **GENERAL**: THIS AGREEMENT SHALL BE CONSTRUED IN ACCORDANCE WITH THE LAWS OF THE UNITED STATES OF AMERICA AND THE STATE OF NEW YORK, APPLICABLE TO CONTRACTS MADE IN NEW YORK, AND SHALL BENEFIT THE COMPANY, ITS AFFILIATES AND ASSIGNEES. This Agreement is the complete and exclusive statement of the agreement between you and the Company and supersedes all proposals, prior agreements, oral or written, and any other communications between you and the company or any of its representatives relating to the subject matter. If you are a U.S. Government user, this Software is licensed with "restricted rights" as set forth in subparagraphs (a)–(d) of the Commercial Computer-Restricted Rights clause at FAR 52.227–19 or in subparagraphs (c)(1)(ii) of the Rights in Technical Data and Computer Software clause at DFARS 252.227–7013, and similar clauses, as applicable.

7. **ACKNOWLEDGEMENT**: YOU ACKNOWLEDGE THAT YOU HAVE READ THIS AGREEMENT, UNDERSTAND IT, AND AGREE TO BE BOUND BY ITS TERMS AND CONDITIONS. YOU ALSO AGREE THAT THIS AGREEMENT IS THE COMPLETE AND EXCLUSIVE STATEMENT OF THE AGREEMENT BETWEEN YOU AND THE COMPANY.